FROM NEURONS TO NOTIONS

FROM NEURONS TO NOTIONS

Brains, Mind and Meaning

Chris Nunn

Floris Books

First published in 2007 by Floris Books

www.florisbooks.co.uk

British Library CIP Data available

ISBN 978-086315-617-5

Produced by Polskabook in Poland

Contents

Acknowledgments

A book like this is the product of a lifetime's assembly of influences and ideas, gifts from other people. I can't remember most of them properly, particularly those likely to have been the most important of all. For instance the maths teacher at my first school had lots of influence on my thinking, all of it good. But I have little idea what he taught me — he certainly never got me to actually do maths particularly well! He was a Viennese Jew perpetually shrouded in smoke from roll-your-own cigarettes, a refugee from the Third Reich known only as 'Mr Philip.' William Sargant, a well known psychiatrist in his day, impressed me greatly when I was at medical school. Subsequently, I was lucky enough to work for James Gibbons and David Kay for a time, and have always been grateful for the example of an intellectual lucidity which I have tried, and failed, to emulate.

To name a few more recent influences, I'm very grateful to Harald Atmanspacher for his discussion of archetypes and for getting me to read some of Hans Primas' papers; also to Erhard Bieberich for his enthusiasm for fractals and for showing what might be done with them in neurology. There are many other people whom I should mention, but most of them are acknowledged in the text so I won't list them here. Two exceptions are Stanley Klein and Jack Sarfatti whom I should like to thank for their stimulating, but totally opposed, views on what can and can't be done with quantum theory.

Embarking on even a short book is not something to be undertaken lightly, for it takes a big chunk out of one's life and is bound to cause heartache of various sorts. So I'm not sure whether to be grateful to Peter Henningsen. His wonderful insights into brain attractor dynamics were what crystallized my own much vaguer thoughts on the subject and provided the spur to write. But at least I can whole-heartedly thank him for helpful comments on my sometimes botched attempts, in Chapter 3, to explain his ideas to a general readership.

Thanks also to the editor of the *Journal of Consciousness Studies* for allowing me to quote extensively (in Chapter 11) from one of the papers published there.

Introduction

Who made your maker? If Self-made,
why fare so far to fare the worse?
Sufficeth not a world of worlds,
a self-made chain of universe?

From The Kasidah of Haji Abdu el-Yezdi,
translated by Sir Richard Burton

The idea that we are nothing more than very complicated machines has been around for quite a while. Some Greek philosophers played with the notion but take-off did not occur until the mid-eighteenth century, since when it has sometimes seemed unstoppable. It was an article of faith for most communists and still is for many in the mainstream of our own societies, especially scientists. After all, regarding bodies as machines has allowed huge breakthroughs in human biology and in medicine. We all have reason to be grateful to this attitude, if only for decent pain relief when we go to the dentist. But, despite its obvious value in relation to our bodies, I think it has only limited validity in relation to our minds.

What can take the place of the machine metaphor? Machines are understandable and thinking of our minds as mechanisms offers the hope that we may one day discover what makes them tick. Sigmund Freud, for instance, pictured us as a bit like the steam engines of his time — the engine of our conscious selves powered by unconscious boilers fuelled with sexual energy (he called it 'libido'), possessing a 'superego' like the steam regulator of an engine. The Behaviourists who ruled academic psychology fifty years ago thought of us as 'black boxes,' with inputs and outputs. More recently, many were seduced by analogies between brains and computers. None of these analogies, however, proved adequate. That leaves us with the question; how should we understand ourselves and our worlds of sensation, thought, feeling and relationship?

In a recent book *(De la Mettrie's Ghost),* I argued that our minds are actually more like stories than like machines. Consciousness helps to select and edit what gets into our memories, I said, and what's in memory is responsible for our choices and actions. Indeed, there's a sense in which we *are* our memories, accumulated, nurtured and pruned over a lifetime. It is a picture, albeit one with lots of complications, that offers our conscious selves a degree of freedom from neural determinism — freedom from the 'brain machine' in other words — and also from the even tighter coils of social determinism. 'Free will lives!' was the message. Seeing exactly how it could live, without throwing out the baby with the bathwater (that is, without rejecting what is true about neural and social determinism), involved a bit of explaining. However, the basic argument was straightforward enough and, above all, was consistent with what we know about brains and human nature.

The main problem with it was that the whole picture focussed too much on the machinery of memory, decisions and the like, without putting them into context. What is the stuff of memory? What choices are on offer out there in the world? I had left out most of the background that's needed in order to properly understand minds, especially conscious minds. The book gave an accurate snapshot of what we are like, I believe, but people might rightly ask for the full movie. After all, our minds live in bodies, they depend on relationships with the world around, especially other people, and they swim in a sea of culture. How can one possibly get to grips with all this complexity? Well, it is in fact possible to make a start on doing so. And an especially promising approach involves picturing brains, minds and the societies in which they exist as landscapes of a special sort: not landscapes as in an oil painting, but more like cloudscapes in a storm, constantly heaving, changing, billowing and flowing. The connection with the stories and memories of my earlier account is down to the fact that these memories, and the story lines existing in society, can be pictured as moulding landscape features.

I'm asking you to join me on a journey through a whole range of landscapes — biological, neural, social and more — which will take us from early views of archetypes, through the mathematics behind neural network behaviour (as briefly as possible, you may be glad to hear!), on to memes, and then far beyond to the nature of time itself. On the way, we shall think about the purposes of sleep, the experience of dying and much else. It's something of a roller-coaster ride. It has to be like this in order adequately to reflect the quicksilver dynamism of our minds.

By the way, this is a 'stand-alone' account. As writers of detective story series like to say: 'You don't need to have read the first volume in order to enjoy this one.' Though based on some of the same science and ideas as the previous one, it is less technical but quite a bit closer to the full story: a tale that is gradually unfolding as we discover ever more about brains, minds and the material world.

1. Archetypes, Attractors, People and Rabbits

Sir Richard Owen was probably, for many years, pleased with his life in London. 'A tall man with great glittering eyes' in the somewhat over-generous view of a famous contemporary, the historian Thomas Carlisle, he was by 1850 widely acknowledged the greatest zoologist in the most remarkable society that the world had ever seen. Tutor in zoology to some of Queen Victoria's children among many distinctions, he was also prime mover behind plans for the marvellous Natural History museum in South Kensington. This finally opened its doors in 1881, eleven years before his death. Yet, by the time he died, his reputation was in shreds. It seemed fitting to some that his commemorative statue in the museum was of blackened bronze, standing close to Charles Darwin's shining white marble. For Owen had tenaciously resisted notions of evolution and had in consequence been successfully savaged by the biologist Thomas Huxley *aka* 'Darwin's bulldog.'

In earlier days, Owen had developed the notion that the body plans of all animals are expressions of 'archetypes' held in the mind of God. Therefore, so he supposed, evolution was out of the question. Whatever his subsequent doubts — and he must have had doubts, since Huxley convincingly demonstrated that some of the 'evidence' on which he had based his theory was faulty — he never acknowledged that he might be wrong. Darwin, in uncharacteristically catty vein, attributed this recalcitrance to Owen's jealousy at his (Darwin's) success. However, as it turns out, Owen was in a sense quite right; not about 'no evolution' but about the usefulness in biology of the idea of archetypes. In fact the concept is useful not only in biology, but also in psychology and sociology.

'Nonsense!' many will cry. Nevertheless this book aims to show that the claim about archetypes is not nonsense, though they generally go by a different name these days. The idea behind them is in many respects equivalent to the notion of the 'attractors' that play a central part in chaos and complexity theory. The fact that Owen can now be regarded as having had a (partially) valid point of view is down to some of the latest discoveries in molecular biology.

It was generally supposed, twenty or thirty years ago, that the genetic code precisely specifies which particular proteins a cell will synthesize, plus the sequence and quantity of synthesis, which in turn exactly determines what an animal will be like — its 'phenotype' as the jargon has it. A few biologists had developed more sophisticated views, but they were in a minority. Imagine most scientists' surprise therefore when it was subsequently found, thanks to all the work put into deciphering genomes, that whole chunks of code could vary between individuals, or even be completely missing, with no apparent effect on the phenotype. Of course this isn't always true. Even quite tiny variations in some genes can have dramatic effects, as manifest in many hereditary diseases. But the fact that there is no discernible effect *sometimes* means that it must be wrong to suppose there is always a one-to-one relationship between genes (genotype) and bodies (phenotype).

Various attempts were made to explain the puzzle. Some suggested that mothers kick-start development in the right direction, through programmes that they implant in their eggs. Such programmes do exist, but it is hard to see how they could solve the puzzle on their own — at least as far as complicated animals like us are concerned. Others relied on the idea that most of our genome consists of useless 'junk' that has no function. This was never very believable, despite all the 'selfish gene' rationalizations that were used to lend it plausibility. Nowadays it is realized that much of the so-called 'junk' in fact has useful or essential regulatory and stabilizing functions. Then people claimed that there is a lot of redundancy in the system so that if one part of the code is abnormal and tending to cause phenotypic deviance, other parts can often take over and compensate, so resulting in a normal body after all. There's a lot of truth to this latter explanation but it has a sort of *ad hoc* quality, which suggested that an essential explanatory concept was still missing. After all, how could the system 'know' what is the proper course of development, so as to allow genes to 'take over' when necessary from deviant partners?

The missing concept can be supplied by the mathematical concept of an 'attractor space': a space which is structured by all the possibilities inherent in the genetic code. The attractors in question are the possible phenotypes to which genetic codes could give rise in the course of the hugely complex processes of growth and development. For instance some particular sub-set of genes might allow a wing to develop. These genes could thus be regarded as having added a 'wing attractor' to the imaginary space. The big advantage of using this notion is that it decouples genes from their phenotypes to a degree. We no longer have to think of genes as being like the instructions in a mechanical loom, which inevitably and directly produce a particular pattern in the weave. Genes are in fact more like a guiding hand behind huge complexities that have their own dynamic and their own momentum. In consequence, it is possible to imagine that differing sub-sets of genes could add the *same* attractor to 'attractor space.' Hence we need no longer be surprised by the fact that seemingly identical phenotypes can sometimes arise from dissimilar genetic codes.

Conrad Waddington, a biologist who specialized in studying embryos, suggested an equivalent notion fifty years ago. He proposed the idea of what he called an 'epigenetic landscape,' created by genes, which consists of hills and valleys down which the ball of development can roll. Changes in individual genes might affect the steepness of slopes or the depth of valleys, but would not necessarily alter the overall topography — and hence need not affect the course of development. His ideas, while much admired, had surprisingly little general influence at first. Perhaps they were too far ahead of their time, which was dominated by the much more reductionist stirrings of molecular biology and what subsequently became 'neo-Darwinism.'

It is surmised that this possibility underpins at least some major evolutionary changes. And major change has always, *pace* Richard Dawkins, presented conceptual conundrums. Despite the plausible 'Just So' stories provided by Dawkins and fellow neo-Darwinists, it has never seemed very believable that a change from walking on all-fours to walking upright, for example, could happen by huge numbers of tiny incremental steps, since some at least of such steps would surely have proved detrimental purely by the laws of chance, thus halting the process of change. The attractor space idea, however, allows variations in genotype having no discernible effect to accumulate until a threshold is passed and the space suddenly flips, resulting in the appearance of a totally new phenotype (which may very occasionally be an improvement on the old, and thus able to win out in Darwinian competition). It is not the whole story about big changes as it seems that they can also occur as a result of mergers between wholly separate genomes; as when our primordial, single-cell ancestors incorporated the bacteria that subsequently became our mitochondria, the power houses of our cells. Nevertheless slow accumulation of genetic change, followed by sudden flip in phenotype, is probably a more frequent process than merger.

In the rest of this book, we shall mainly be thinking about a range of other landscapes and attractors; not ones based on genes (though these will crop up again from time to time) but on the behaviour of neurons in our brains or 'stories' in our societies. The genetic example, however, helps to show not only what sort of thing attractors are, but also how Owen was right. There are indeed biological forms inherent in the nature of things, which thus correspond to his 'archetypes.' They may not exist (or, on the contrary, they may exist for all we know) in the mind of God as Owen supposed, but they are certainly embodied in the cascades of complexity that underpin growth.

Mental archetypes

Carl Gustav Jung, unlike Richard Owen, died with his reputation largely intact. He too is well known for his concept of 'archetypes.' What are these entities? The answer to this question is often considered obscure. It was not for nothing that Jung was the son of a Swiss pastor. He was something of a mystic and wizard as well as a widely admired guru, who as a youngish man held conversations with his own spirit guide, an entity named Philemon. Later on, he developed an obsessive interest in alchemy. Not only was his thinking intrinsically obscure at times, but he also liked to make it appear more difficult than it really was: 'That is why I prefer ambiguous language, since it does equal justice to the subjectivity of the archetypal idea and to the autonomy of the archetype,' he wrote in one of his letters (dated 17/6/52).

Another problem was that mysticism sometimes influenced his picture of archetypes. During one of the spirit guide conversations, he reported that Philemon viewed thoughts as having an independent existence like animals in the forest, or people in a room, or birds in the air, while Jung himself naturally treated them as his own productions. Philemon's opinion seems to have stuck in Jung's mind and attached to his view of archetypes. Before 1919, he had called them 'primordial images.' What he meant by this was that they were recurrent themes that cropped up in all human cultures. If he had stuck with this earlier term, people might have understood him better. Favourite examples that he used included 'The Mother' and 'The Mandala' (a symbol of some sort enclosed within a border, such as a cross within a circle). Crucially, for our purposes, he also made a distinction between the expression of such themes in consciousness, or in works of art or craft, and the *predisposition* to express them. He referred to the former as an 'archetypal representation' and to the predisposition as the archetype itself: '... in itself the archetype is an irrepresentable configuration whose existence can be established empirically in a multitude of forms. The archetype of the "mother," for instance, manifests itself in infinitely many forms and yet the one common characteristic of the mother-idea always remains intact,' he wrote in another letter (11/6/58).

Jung's idea of an archetype, stripped of the flummery with which he coated it, thus appears equivalent to the modern notion of an attractor. This correspondence accounts for much of the confusion of his descriptions. At one time, he described his archetypes as existing in a sort of quasi-mathematical, Platonic realm. Then he changed his mind and conceived of them as emergent properties of human biology. Then he kind of half rejected this idea.[1] In fact, of course, attractors are both mathematical concepts and emergent properties of complex systems like human brains. Jung was groping towards the concept, but did not have it any clear form. Like Owen, though, he did have a valid point.

Freeman's rabbits and attractors

So the concept of archetypes has a longish history in biology and psychology (of course it has an even longer one in philosophy, dating back to Plato and earlier, but I won't be discussing that here). According to Owen, archetypes have a big part to play in our physical make-up, while Jung gave them a central role in our psyches and also in influencing the art that we produce. How has the notion fared more recently? As we have seen it has much in common with the idea of attractors, which occupy centre stage nowadays; it's time to look at an explicit example of a possible role for them.

Walter J. Freeman is a well-known neuroscientist and something of a polymath, whose main base is in Berkeley, California. His colleague Christine Skarda and he published a particularly famous paper in 1987. It was entitled *How Brains make Chaos in Order to Make Sense of the World.* They, their results and the rabbits who were the subjects of their research, were discussed worldwide. Their experiments involved EEG (electroencephalography), and I had better give you some background on that before describing their findings and conclusions.

Back in Owen's day, it had been noted that brains seem to produce tiny electrical currents. However, nothing much was done with this observation at first because the currents are so tiny that Victorian technology was not up to measuring them reliably. It was forgotten for a time, then rediscovered by Hans Berger in the 1920s, using better technology. He was the first to describe the so-called 'alpha rhythm' (the rhythmic 8–13 cycle per second electrical changes that can be detected over the back of people's heads when their eyes are closed). His finding was taken up and developed by Lord Adrian in Cambridge, the centre of the scientific universe at that time, and has been used in research and for clinical purposes ever since. Because there were a lot of epileptic patients in Zambia, for instance, I imported an EEG machine when working there in the late 60s. It provided an object lesson in just how small are the brain's electrical currents. To start with, it showed only mains frequency activity that patients were picking up from nearby wiring, which completely swamped what they were producing themselves. They were behaving like radio aerials. We got round the problem by burying an old radiator from a lorry outside the office and earthing patients and EEG machine to that. Then the dry season came, the water table fell and the earth no longer earthed. Each recording session had to be preceded by copious libations of salted water poured on the radiator's grave!

EEGs use quite big electrodes, usually placed on the scalp but, if you can put a tiny electrode in the brain, you can also record the action potentials of single nerve cells. It's possible to do this in animals and also, very occasionally, in people having operations for some disease of the brain. The action potentials occur in a sort of 'sputtery' fashion, sometimes at higher frequency, sometimes at lower. If you were to make separate recordings from just a few such cells, you would never guess that their bursts of activity could add up into the wave patterns seen in EEGs, which are responsive to large numbers of cells and represent a statistical average of aspects of their activity. EEGs, in other words, are representations of the dynamics of electrical field changes in brain systems much larger than individual nerve cells.

When people or animals are awake, their EEGs often have a 'fractal' structure.[2] The changes they show tend to look much the same whether you record from a large area or a small one, whether you focus on longer time periods or shorter ones. Dynamic systems of this sort are often said to be 'poised on the edge of chaos,' meaning that they can shoot off down all sorts of unpredictable routes, which look random to an observer though in fact they are deterministic. The relevance of all this here is that the states these systems end up in can be regarded as attractors, which

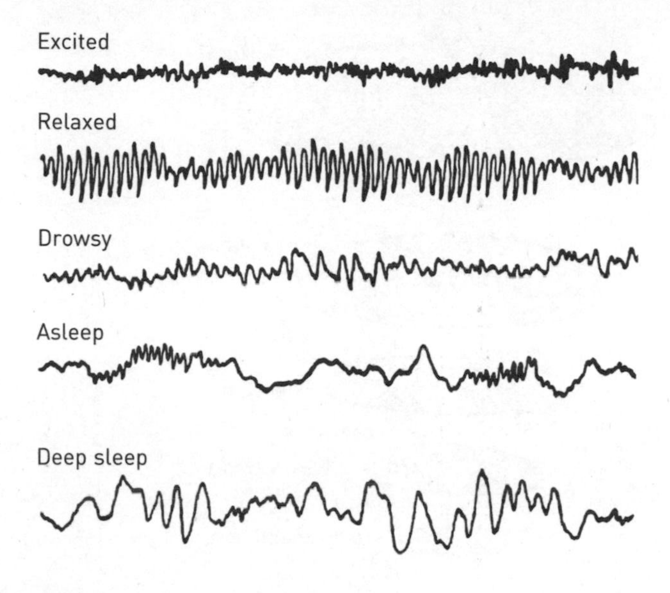

Figure 1: Normal EEG

appear to constrain the routes that the systems are able to follow. So back to Skarda, Freeman and the rabbits.

What the rabbits did was to sniff a variety of odorants provided by the researchers. What Skarda and Freeman did was to examine the patterns of chaotic EEG activity in the rabbits' olfactory lobes while they were sniffing. They saw that different odorants gave rise to different patterns. Giving the same odorant again did not produce exactly the same pattern as on the first occasion; but there were fewer differences than those seen in response to a dissimilar odorant. Indeed, the patterns seen depended not only on what chemical the rabbit was sniffing but also on things like how aroused it was and what sort of training it had undergone in the past.

In a 1999 book *(How Brains Make up their Minds),* Freeman described what these results, and others from related experiments, might mean. He thinks that what went on in the rabbits is best pictured in terms of an imaginary landscape containing many attractors, which shape the dynamics of the electrical activity in their olfactory lobes. 'A new odorant is learned by adding a new attractor with its basin, but, unlike a fixed computer memory, an attractor landscape is flexible.' he wrote. 'The attractors are not shaped by stimuli [such as particular odorants] directly,' he pointed out, 'but by previous experience with those stimuli ...'

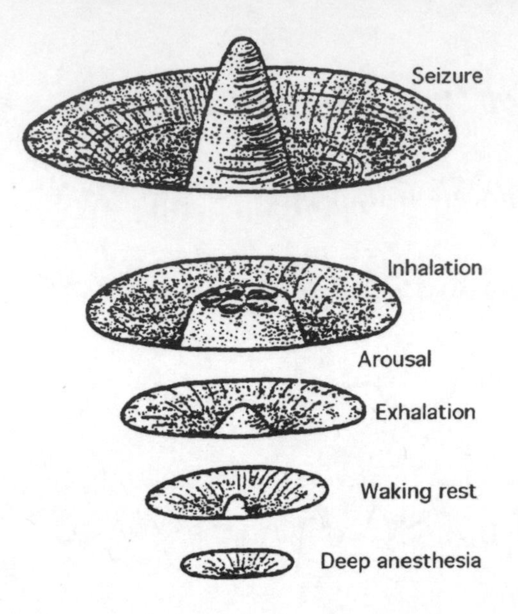

Figure 2: A simplified diagram of attractor landscapes in a rabbit olfactory lobe. The landscape changes according to how the rabbit is breathing and what it is doing.[3]

He was saying that the rabbits' memories of smells, triggered by their current sensations, were equivalent to the attractors that he and Skarda saw in their brains. If the creatures were conscious, as seems likely, then the flow of their conscious experience of smells corresponded to the dynamic determined by their attractors. It seemed natural to Freeman to suppose that the same might be true of all perception, and perhaps of the flow of conscious experience in general, not only in rabbits but also in people. This was such a fascinating, profound and far-reaching suggestion that one might have expected others to go overboard for it. On the whole, they haven't. I don't want to give the impression that there is no interest in this line of research nowadays. As we shall see in due course there's plenty going on, but it seems to be regarded as a bit peripheral to mainstream concerns and we need to look at some of the reasons why this should be so. There are four main ones. Two of these have to do with what might be termed scientific fashion, the third has to do with practicalities, while the fourth is seen by many as a more fundamental obstacle to Freeman's proposals. Let's take them in this order.

Nowadays people tend to think of MRI (Magnetic Resonance Imaging) as *the* most useful technique in whole-brain research. It is certainly wonderful, almost magical technology. However, it does have drawbacks, while the relationship of what is measured by the most commonly used MRI technique (fMRI) to what is going on in nerve cells proved difficult to ferret out. In fMRI what's measured is differences between the amount of oxygen in the blood in different parts of the brain. When nerve cells are particularly active, they naturally use more oxygen and their local blood vessels compensate by supplying them with extra oxygenated blood. In fact the blood vessels slightly overcompensate, so the oxygen levels in more active areas actually increase, which is the reverse of what you might expect. But what aspect of nerve cell activity is reflected in this increase in oxygen level? Using marvelous experimental techniques, Nikos Logothetis finally pinned down an answer to this question. Prior to his work most people had expected the answer to be something like 'frequency of the action potentials produced by nerve cells.' As it turns out, the answer is much closer to 'local electrical field changes.' This means that fMRI is indirectly measuring much the same variable as EEG, which is a direct consequence mainly of local field changes though also, to a lesser extent, current flows along bundles of nerve fibres.[4] The advantage that fMRI has over EEG is that its spatial resolution is much better. It can pin down changes in a few cubic millimeters of brain, whereas EEG (when recorded from scalp electrodes, which is the normal method) is accurate only to several cubic centimeters at best. On the other hand, EEG has a much better temporal resolution, down to one or two hundredths of a second, while even the most powerful fMRI scanners take about a second for each measurement. The machines can record images faster than this, but there's no point in doing so because the bloodflow changes that they measure take a second or two to show up.

So EEG studies are more relevant to questions about short-lived changes in the brain than are fMRI ones. The technique may be a bit out of fashion at present, but is not at all out-of-date and is still being refined. Moreover, Skarda and Freeman got round the poor spatial resolution of scalp EEGs by recording direct from the surfaces of a part of rabbit's brains (that is, their olfactory lobes, which are responsible for their ability to smell things).

There *has* been enormous enthusiasm for fMRI studies of higher brain functions, which have produced a string of fascinating discoveries. But, if you spend several million dollars getting hold of a scanner and all the ancillary facilities that are needed, you're going to have to use it and you're not going to have much attention to spare for anything else. As was pointed out above, fMRI techniques are inherently incapable of detecting the sort of rapid dynamics that Freeman was interested in. It's just too slow.

Another long-lasting enthusiasm has been for laboratory studies of the activity of single nerve cells or very small numbers of them. These experiments, too, have produced a seemingly endless series of discoveries, starting with the finding that there are cells in the visual centres of cats that respond to lines shown to the cats only if they (that is, the lines) have some particular orientation; then there was all the work, which is still continuing, on cells (in what's termed the hippocampus) of rats that 'recognize' when their owner reaches some particular place in a maze; more recently there was the discovery of 'mirror neurons' that are active both when you are performing some particular action and when you are simply watching someone else do it. The whole field has been almost dazzling in its productivity. But you can't see Freeman's dynamics when you are looking at single cells. It's just not possible.

Those are two main reasons why current mainstream research is largely blind to Freeman. It simply can't 'see' what he looked at. Then there is the question of complexity. Rabbit olfactory lobes are simple compared to, say, the visual system of monkeys. But the latter is what most researchers interested in perception like to focus on, for the very good reason that an enormous amount is already known about it, and it's usually better to build on what is known rather than start from scratch somewhere else. No-one, not even Freeman himself, has been able to cope satisfactorily with the extra complexity involved in trying to do studies equivalent to his rabbit ones on monkey vision. It would be difficult, to put it mildly, to develop the techniques needed to discover whether what's true of rabbit olfaction really does apply more generally — and while there is so much exciting stuff being produced by fMRI and single neuron studies, there's not any great incentive to put a lot of effort into what might turn out to be a blind alley.

These problems relate to practicalities only, and could be resolved sooner or later. Indeed the situation may be changing already. There's a new technique involving dyes which change colour according to the activity of nerve cells. Its temporal resolution is as good as that of EEG and its spatial resolution even better than fMRI. It seems ideally adapted to the search for chaotic neural dynamics, but of course there's a drawback. It can only be employed in anaesthetized animals at present, so cannot readily be used to investigate Freeman's ideas about conscious ones. But maybe in the future ways will be found to apply it more widely.

However, there is a more serious obstacle to research than these technical ones. Some think it likely that Freeman's whole approach is a will o' the wisp, and thus inherently not worth pursuing. 'Attractors' seem to these objectors to be simply conceptual conveniences, which exist only in the imaginations of mathematicians. Suppose a meteorologist interested in clouds, for instance, were to develop an idea of 'castleness.' 'Ah, that cloud's a definite Norman,' he might say. 'Look, there's a Windsor over there, and I think that one's probably a Caernarfon.' Well, maybe a concept like this would help him to think about clouds, and might even reflect aspects of their actual structure in some way. But if he started to claim that the castles had real, independent existence of some sort, he would probably soon be off to the nearest psychiatric institution.

Neuroscientists tend to be suspicious of apparent abstractions like attractors. They prefer to deal with what they regard as 'real' objects, such as pointer movements caused by flows of electric charge or blood flow changes detected by MRI. They want to study ion flows and cell connections, not anything that seems too airy-fairy. And Freeman can be read as suggesting that the flow of conscious experience, which is the most 'real' yet elusive thing in the world to each of us, is an expression of attractor dynamics, a concept that appears even more abstract and fanciful than that of castles in the clouds. 'There must be something inherently wrong with the man's picture,' is the mainly unspoken conclusion, 'so it should not be taken too seriously.' I believe that any such conclusion is wholly mistaken, and hope to show you why I think so as the book unfolds.

Just to get in an initial dig at the conclusion, let's take a look at a concept that underlies much of what neuroscientists do like to study, namely that of the electron. 'That's not a concept, it's a particle!' you may say. But is it? Physics shows that an electron is certainly not a particle in the same sense as we take a grain of sand to be a particle, for it can also behave like a wave. It carries 'charge,' whatever that may be, and the even more mysterious 'spin' (which is not at all like the spin of a cricket ball because it comes in discrete units and is present in any direction that you care to measure). In fact the best verbal descriptions of an electron produce only paradox and apparent nonsense. The only good description we have is mathematical — Paul Dirac's equation of the electron. And that seems to show that it does not have what one might call 'real' existence at all; it is a matrix of potentialities that manifest actuality only in the context of particular observations or experiments. Despite this, all neuroscientists would agree that electrons appear to possess objective reality and have properties essential to our very existence, as well as being responsible for determining the results of their experiments.

If I suggest that attractors may be just as 'real' in relation to complex systems as are electrons in relation to the basic properties of matter, I am not necessarily headed for cloud-cuckoo land. In fact I would simply be following in the footsteps of Owen or Jung, both of whom saw the entities that they proposed as having some sort of quasi-independent existence and being in a sense fundamental, in much the same way as electrons are viewed as fundamental in physics. Attractors may be products of all the causes operative in dynamic systems, but then electrons appear to be products of whatever causes have produced that extremely mysterious entity, matter. The next step, therefore, is to take a look at what mathematicians mean when they talk about attractors. Unsurprisingly, doing this will make the whole concept look very abstract indeed, and could reinforce your darkest suspicions about them being mere castles in the clouds. Subsequent chapters will be about bringing the concept down to earth again in the context of brains, psychology, social behaviour and beyond.

2. The Laws of Attraction

I had better come clean straight away and confess that attractors exist in a wholly imaginary space, according to mathematicians. It's called 'state space.'

'So what's the point of reading any further?' I can hear you mutter, 'They *must* be just figments of some fevered mathematician's imagination.' Well, state space itself was a product of the imagination of Henri Poincaré, who died in 1912 but has been remembered ever since for the usefulness and penetration of his mathematical insights. He could easily have pipped Einstein to the post in developing special relativity, it has been claimed. Henri himself was no fantasist, though one of his cousins may have been. This was Raymond Poincaré, a politician who was president of France at the time of World War 1.

In any case, there's nothing wrong with picturing things as existing in an imaginary space, for we all exist in exactly that way ourselves. Ultimately, we are all composed of the particles described by quantum theory, which 'exist' according to physicists in an equally imaginary space called 'Hilbert space' — after the famous German mathematician David Hilbert, who was a younger contemporary of Poincaré. I suppose the difference between us and attractors is that we exist in what we take to be 'real' three-dimensional space as well as in imaginary space, while attractors manifest in 'real' space only as the behaviour of particular dynamical systems. You could not go into a supermarket and pick a dozen of them off the shelf.

The basic idea behind 'state space' is very simple. It is an arena within which any possible state of a dynamic system can be represented by a single point. Let's start with an uncluttered example: a weight on a spring, bouncing up and down in a perfectly straight line. Suppose, further, that you want only to represent the possible momentary positions of the weight in your imaginary space. The weight might be at the top of its bounce, or the bottom, or any point in between. Those would be its momentary states. So the corresponding state space needed to show its possible positions would be nothing more than a single line. Put a dot anywhere on that line and it would stand for some particular state of the system.

If the weight can swing a bit from left to right as it bounces, then you would need a two-dimensional space to allow representation of all possible positional states of the system with only a single point; one dimension for its degree of bounce and another for the degree of left-rightness. A sheet of paper would do just fine for this. If the weight could also swing backwards and forwards as well as left and right, you'd need a third dimension and thus a cube for the state space.

As the weight bobs up and down (let's suppose for the moment that it is confined to up and down only) it will of course steadily lose amplitude (that is, the bobs will get steadily smaller) due to factors like energy dissipation in the spring and air resistance. Sooner or later it will come to rest at a point determined by the inherent length of the spring and how much the weight stretches it. This point will also be represented in the corresponding state space (the single line in my original, pruned down, example) by another point. And the latter point is called an attractor because, wherever the system starts off, it eventually ends up there. As you would expect, it is called a 'point attractor'; all very simple, indeed almost banal so far. Mathematics, unfortunately, has a nasty habit of starting off simple and getting complicated very quickly. Its great virtue, however, is to show how simplicities are connected to complexities; what patterns underlie the diversity of phenomena that mathematicians themselves come up with, or that we can see in nature.

If we now allow our weight to move sideways and to and fro, as well as up and down, it has three 'degrees of freedom,' as they are called, so a cubic state space is needed to show its positions. If you start the system off at any point in its state space that is away from its 'attractor,' it will trace a path through the space that will end up at the attractor. Let's say you start it with the spring fully stretched and held out at right angles, then it will accelerate fast — maybe straight towards your face, so watch out! — and the corresponding points of its momentary positions in state space will also rapidly move. Gradually, however, it will calm down and finally end up hanging motionless. So the plot of its successive positions in state space will be like an inverted cone. The point of the cone is the attractor for this system; the cone itself the 'attractor landscape' — not a very interesting one in this case. However, if you have a complex system containing many attractors, the corresponding 'landscape' can easily look just as complex as some scene from the Scottish Highlands.

A mathematician, who kindly checked on an early draft of this chapter for me, pointed out that what I've written so far is quite misleading, since one can legitimately talk of attractors only in the context of what he termed a 'complete' description of a system. Looked at from one point of view his comment was entirely correct, but from another it was way off the mark. He was correct in that it is nonsense to think of a dynamic system like our spring and weight solely in terms of momentary positions. You have to consider movement as well (technically its 'impulse'). To represent that, for a simple system like ours, you will generally need another three dimensions of state space. I saved this information till now, as I wanted to convey *some* impression of an imaginable attractor landscape — but not *too concrete* an impression, which is why I have not provided nice diagrams of pendulums and cones. A six-dimensional cone, the simplest possible attractor landscape, is definitely not imaginable or drawable! All the same such objects do in a sense exist, if only in the minds of Poincaré and his successors, and are analogous to ordinary cones.

When the mathematician said 'complete,' on the other hand, he could be regarded as misleading in the sense that a merely six dimensional description of a weight on its spring would be far from complete. For completeness you would need an extra dimension for the changes in spring elasticity caused by temperature variations, yet more for electromagnetic effects due to movement through the earth's magnetic field, a huge number to describe the consequences of any atmospheric turbulence caused by the movement, and so on and so forth. So you do not in fact need a realistically 'complete' description of a system in order to introduce the idea of attractors; all you need is a mathematician's simplified, idealized 'completeness,' because additional factors, like turbulence and so forth, will be represented all right in 6-D space as jitters occurring as the point moves through the space.

Dynamic systems that don't have an energy input have point attractors, as does our weight on its spring. This is because they sooner or later dissipate energy and end up in an inert state. But most systems of interest to us, brains and the like, *do* have energy inputs. If sufficiently complex, they will probably contain point-like attractors, but are also liable to harbour the two other types, which I have not yet described.

If, instead of holding your weight on its spring steady, you put energy into the system by moving your hand from side to side, its final state may be to end up swinging gently in a circle. Whether or not it does so will depend on the frequency and regularity of your hand movements, plus system variables such as the length of the spring and the strength of gravity. The corresponding attractor, instead of being a point, will now be a 1-D line, called a periodic attractor. Alter how you swing it a bit, and you may end up with the weight both moving in a circle and gently bobbing up and down, in which case the system embodies a 2-D, 'quasi-periodic' attractor, but we'll lump this together with the ordinary periodic attractor for the purposes of this book.

If you pump enough energy into the system, by moving your hand ever more quickly, the weight will fly off all over the place quite unpredictably and will never settle down to any obvious regular behaviour, though its actions will always be 'bounded' (unless the spring breaks) because they will always be confined within the arena determined by the maximum length of the spring. You'll have turned your system either into a chaotic one, or into a close approximation to chaos (it may still contain *some* regularities). However, it will still contain attractors — so-called 'strange' ones — which will themselves change with the amount and direction of the energy you're pumping in. The type of attractor that a system will harbour, point, periodic or strange, can be determined precisely, at least for simple assemblies like our weight and spring, by numbers called 'Lyapunov exponents' after Aleksandr Lyapunov, a Russian mathematician who described them in his 1892 doctoral thesis.[1]

Most people have seen the simplified diagram of a strange attractor that was originally discovered by meteorologist Edward Lorenz, one of the founders of chaos theory. It resembles a butterfly — two large wings and a narrow connecting bit — which is very apt, given the famed 'butterfly effect' of meteorology. A butterfly flapping its wings in West Africa may cause a hurricane to hit Florida was a favourite image at one

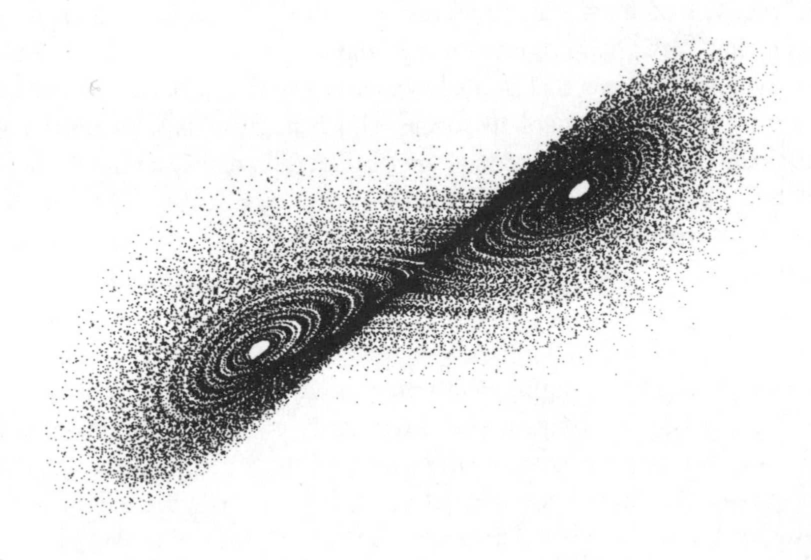

Figure 3: The Lorenz 'mask,' a three-dimensional attractor discovered by Edward Lorenz in 1963 while modelling weather patterns. It was the first known instance of a chaotic attractor.

time; the butterfly is now sometimes sited in Hong Kong, which change may itself be a chaotic phenomenon, reflecting perception of the ever-increasing importance of China. Lorenz himself had it in Brazil, causing a tornado in Texas.

Lorenz's picture is actually a representation of the dynamic trajectories that a particular chaotic system can follow. What's specially interesting about it is that you would have to pick a starting point with infinite accuracy, if it happened to be somewhere in the narrow bit between the two wings or round the edges of a wing, in order to say where the trajectory will go from there. The most central parts of each wing tend to behave like point or periodic attractors in the short or medium term, though in the long term they are just as 'strange' as the rest. That degree of accuracy is of course inherently unattainable. In general, if you were out by even an infinitesimal amount in describing the current state of your system, it would end up somewhere totally unpredictable. All you could predict is that it would be *somewhere* within the spectrum of states described by the strange attractor. By the way, not all strange attractors have the same shape as the Lorenz one. They are often a lot more complicated and can take all sorts of, frequently beautiful, forms (see Figs.4–6).

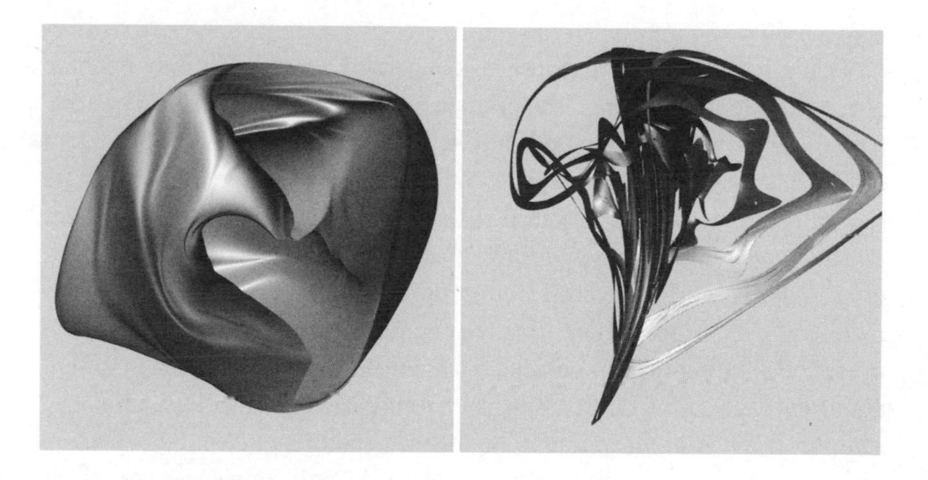

Figures 4 and 5 (Valery Tenyotkin)

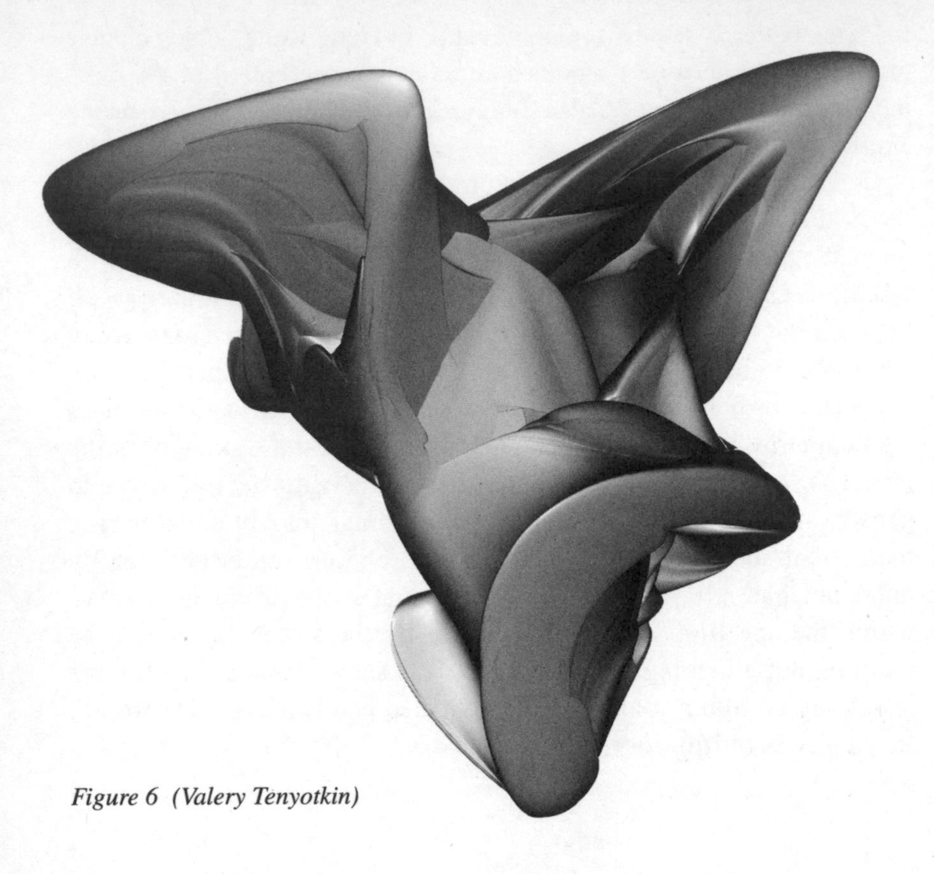

Figure 6 (Valery Tenyotkin)

That's why climatologists are so concerned about what may become of us with global warming. The climate is thought to be a chaotic system. Extra energy is being pumped into it, and it is only too likely to zoom off down some unpredictable path. We may or may not get a thirty metre rise in sea levels due to Antarctica thawing out. We could even trigger an ice age rather than get warming, as a result of unanticipated feedback effects. There's a lot of effort going into trying to find out what the climate was like, especially at times in the past when there was more of a greenhouse effect than at present. You can discover a surprising amount about average temperatures and even rainfall long ago, by examining samples taken from deep in the ice or from sediments. What's happened before is, perhaps, likely to happen again, but of course we can never know for sure because some of the energy is being injected in different ways now from what occurred before. For example there were no jet contrails in Jurassic times to contribute to the greenhouse. And a difference like this might, or might not, affect the climate's future.

As mentioned earlier, really complex systems, brains for instance, contain a whole range of attractors often of all three types. Their attractor landscapes are correspondingly complex. And each will have its own little kingdom. Any line in state space that strays over the border will become subject to the rule of that attractor, and its allegiance to others will be severed. This is what Walter Freeman meant when he wrote about a 'basin' of attraction. He was picturing the landscape as made up of ridges and valleys, with an attractor sitting at the deepest point of each valley. Any path leading downwards would be within the 'basin' of that particular attractor. In the case of our weight and spring, of course, the whole of its state space comprises a single 'basin,' but things are rarely that simple in the real world. Points in state space represent momentary conditions of a *whole* dynamic system, so these paths are equivalent to descriptions of how the system must evolve.

All dynamic systems, such as brains, contain attractors (point, periodic and/or strange) considered as potentialities in state space. By the way, there's a technical issue that I should tell you about here, even though it is often ignored by neural network theorists. If you map the dynamics of actual live neurons in a brain into a state space, it follows that any so-called 'point' attractors in it are really no more than point-like (strictly speaking they will be quasi-periodic attractors occupying a compact volume of state space). This is because of the fact that 'genuine' point attractors exist in systems lacking an energy input — and any neuron lacking an energy input would have to be dead and deep-frozen!

Be that as it may, each attractor in the brain will have its own little kingdom. But their 'existence' may be very fleeting if the system is not closed — if, in other words, outside influences are affecting it. For such influences are liable to alter the attractors in a system. In the example given earlier, if you have a pendulum whose swings are winding down towards a point attractor, for instance, and then gently and regularly move it, you may convert the former point attractor into a periodic one. But if it is blown about by a gale, a whole series of different strange attractors are likely to occur in very rapid succession. Mathematicians will sometimes say that attractors exist only in closed or nearly closed systems. And it's true that they only manifest in such systems because, when a system is too open to outside influences, they don't 'exist' for long enough to have any observable effect on its dynamics.[2] Put, another way, the dynamics are changing too fast to be usefully described in terms of attractors. Nevertheless they are evanescently there in the form of a rapid succession of what might be termed 'virtual attractors.' Going back to the analogy with archetypes, you could say that attractors themselves are like archetypes, but mathematicians sometimes use the term to refer to the visible manifestations of attractors — their 'archetypal representations' as it were.

We tend to think of landscapes as static, even though we know they are not. Over geological time the bottom of a sea rose up and became the peak of Mont Blanc. Volcanoes erupt and flood huge areas with molten rock. The very continents merge and break apart. The same can apply, generally on a much faster time scale, to attractor landscapes even in relatively 'closed' systems. New attractors are constantly being formed and old ones can be obliterated, as a consequence of evolution in the system altering its dynamics. Whenever a new attractor forms or disappears, the dynamics of the whole system are likely to be affected, and the remaining ones have to adjust. They may become shallower or deeper and the size of their 'basins' is likely to be affected. In fact the attractors themselves can be regarded as having their own dynamic. Although this is dependent on whatever causal effects are occurring in the system, it can nevertheless appear to take on a life of its own when things get sufficiently complex, as they do in human societies for example. We'll be taking a closer look at this topic later on.

Causes and laws

The next question to ask may seem a rather childish one: what are attractors really, really like? We've seen that they are mathematical objects; but they are also embodied, in some sense, in real life systems; they cannot be picked up and handled as separate objects even though they do, as we shall be exploring later, sometimes appear to be in control of their system rather than being simply a property of it. We seem to be no further on, yet, than was Carl Jung puzzling over the nature of his archetypes. He spent much of his quite long life switching from one opinion to another about what they might be, and never getting very far. But we do have a big advantage over him nowadays. He approached the concept from a sociological and mythical angle, whereas we have got there via mathematics, which is inherently far less likely to cause confusion. Jung did realize that his archetypes were 'like' mathematical concepts such as that of the natural numbers, but he could never quite see how to relate this to what was going on in peoples' brains. He tried very hard at one stage, in collaboration with his friend the great physicist Wolfgang Pauli, to ground his concept in quantum theory. In the end, Pauli had to tell him that he was on a wild goose chase as far as the quantum theory of the day (around the 1940s and 50s) was concerned. The same appears to be still true today, though many hope that the situation will soon change.

The answer to the question is actually fairly obvious, provided the maths is taken into account. As suggested at the end of the last chapter, the concept of attractors may have something in common with that of electrons, but they themselves are not all that much like electrons for they have their home only in complex systems, whereas electrons are ubiquitous.[3] Moreover electrons are generally considered to have some sort of 'real' existence that is independent of their mathematical formulation, whereas attractors lack this type of 'common-sense' independence. They are in fact like natural laws, albeit ones that have very restricted and specific applicability. Indeed, if one takes this position, Occam's razor (that is, the logical rule attributed to a medieval philosopher, William of Ockham, which says: 'You shall not multiply entities beyond necessity') suggests that it is preferable to regard them as *being* natural laws. Laws are immaterial, but nevertheless are embodied in material systems and appear to constrain their behaviour, just like attractors.

If this suggestion comes across as fuzzy and out of focus, it is because the whole question of the nature of natural law is similarly hard to get to grips with. Laws appear to constrain and shape our worlds, yet many of them also seem no more than emergent properties of the causes that they themselves underpin. There are self-referential loops involved here, of the sort that always cause confusion. Laws such as those of hydraulics, for instance, are thought to be no more than results of causal interactions between water molecules which are, in principle, determined by deeper laws, ultimately resting on the rules of quantum dynamics. Yet it is impossible in practice to derive hydraulics from quantum dynamics. The laws of hydraulics certainly appear to constrain the behaviour of water molecules, and do so in a manner which leads to their own manifestation. Similarly, attractors are consequences of many layers of lawful behaviour reaching down, in the last analysis, to quantum mechanics [4] or to whatever may one day be discovered to underlie that, but they nevertheless appear to constrain the behaviour of the systems in which they are found.

We tend to think of our most general laws, like conservation of momentum or energy, as truly fundamental. Yet this is not so. Noether's theorem shows beyond reasonable doubt, to use a phrase from a different sort of lawful framework, that they are almost trivial consequences of underlying symmetries.[5] Conservation of momentum, for instance, follows directly from the fact that the laws of physics look the same regardless of the direction in which you carry out experiments. It won't make any difference to your results whether you face north or south, east or west while obtaining them. Similarly, conservation of energy is simply an expression of the fact that experiments will give the same result whether they are performed at lunchtime or tea-time. I know that experiments involving the earth's magnetic field for instance, or ones on peoples' alertness, *would* be affected by these factors, but we're talking general principles here.

Similarly, Einstein's special relativity is a consequence of the fact that how fast you are travelling makes no difference to your perceptions of physical law, while general relativity relies on there being no possible local experiment that could distinguish between being sited in a gravitational field or in an accelerating rocket ship. We don't know why these symmetries should be as they are. Perhaps they too are dependent on some deeper law. In fact the ones relating to general relativity must be so, since they are incompatible in their present form with the laws of quantum theory.

Nevertheless, despite the fact that they are probably not truly fundamental, these laws certainly appear to constrain us and our world. Conservation of energy prevents us from solving all our energy problems by developing perpetual motion machines. Conservation of momentum rules when it comes to car crashes. And the same applies all the way up the hierarchy of natural law. Anyone trying to swim in a mountain stream will be greatly affected by laws of hydraulics, in what seems to the swimmer a most direct and compelling sort of way. Laws of physiology are clearly no more than outcomes of others governing the behaviour of those very complicated biochemical systems, our bodies. Yet we ignore them at our peril.

Attractors, I suggest, are at the top of the hierarchy of natural law. There are many variants, each with very restricted applicability (that is, to some particular type of dynamic system). They are even more restricted than are the laws of hydraulics, which apply only to one particular type of physical system (liquids). All the same, despite their parochial character in this respect, they retain the law-like characteristic of appearing to 'govern' the systems in which they occur. I will be suggesting later on, too, that they may be less parochial than they look at first sight. They may sometimes appear to possess the transcendent quality of natural law: its universal applicability in appropriate circumstances.

'But you've just pointed out that attractor landscapes are often in a state of flux,' someone may object, 'surely that doesn't make them very law-like. Laws are nothing if they are not constant and stable.' Well, even quite simple systems can appear to be ruled by separate laws at different times or in different contexts. An electric motor, for example, is in some respects 'governed' by Ohm's law of current flow and in others by conservation of angular momentum or thermodynamics. So the fact that lots of different 'laws' (that is, different attractors) may come and go within really complex systems does not present any 'in principle' difficulty for the view that they are so law-like as to *be* laws. Indeed, I argued in a previous book *(De la Mettrie's Ghost)* that the ability of our consciousness to modify attractor landscapes may be at the basis of a capacity for genuine free will. Moreover, the fact that there's a dynamic of attractor landscapes, which can be regarded as a dance of laws governing neural and (as we shall see later) social behaviour, has all sorts of fascinating consequences that we'll begin to explore in the next chapter.

3. Attractors in the Brain

If you type the title of this chapter into Google you get, at the time of writing, about 491,000 hits in 0.25 seconds. It's unbelievably neat technology. But making sense of the information, as all of us know, is not nearly as easy as using the search engine. One can readily get nostalgic for the days when it was possible to read up a topic by browsing through a section of a good library, and knowing that you were unlikely to have missed anything important. For instance, if Google included Walter Freeman's book (see Chapter 1) at all, it did not do so where appropriate, which would have been somewhere right at the start. Maybe the book is there down in the Z-list pages, but who has the time or patience to trawl through them? Putting it at the top would have helped people to understand what all the other stuff is about. But at least you can see that the whole field is very much alive, judging by the number of entries it has generated. Let me try to summarize what is going on at present.

It has to be said that the overall impression is that of a science at what might be called an 'eighteenth century' stage of development. There seems to be a lot of faith in some circles that attractors are important, but the overwhelming majority of papers are to do with model systems, some quite artificial, others based on simplified versions of real nervous systems. These papers explore what role attractors *might* play in real brains. Other than Freeman's work, there is almost nothing on what role they *do* play, and even less on the processes of prediction and verification or falsification that one sees in a mature science. Most of the empirical studies relevant to the whole field are about whether or not EEGs are chaotic. These studies are relevant because chaotic systems are liable to harbour strange attractors, and strange attractors behave differently from the other two sorts (point and periodic). There is little agreement even here. It seems that occasionally EEGs are chaotic, more often they nearly are but don't quite make it according to rigorous definition of chaos, and the rest of the time they aren't.

Groups of enthusiasts are still groping towards consolidating the field. One interesting snippet, for example, described how the American 'Human Frontier Science Program' had awarded a research grant in 2004 to three people wanting to study how brains recognize facial expressions — a process likely to involve attractors. One of the researchers was based in Trieste, one in London and the last in Seattle. This geographical spread strongly suggests that the field has not really 'gelled' yet. It's more like what went on, one feels, when Lavoisier was in touch with Priestley about developments and possible experiments to do with the foundations of chemistry.

What sorts of role do people think attractors might play in real brains? The whole topic of neural network theory comes in here. In the early 1980s, John Hopfield of Caltech showed that quite simple networks of (artificial) neurons, having an input and an output layer with reciprocal interconnections, could organize themselves with the help of a bit of training so as to store 'memories' and undertake tasks like pattern recognition. It was hoped that these would lead to huge advances in artificial intelligence. Quite soon, so the fantasy went, we would have networks better able to recognize faces than people can, or performing useful tasks like translating Russian scientific papers into English. The military, too, got very excited about possibilities for getting machines to recognize camouflaged tanks and the like. Alas, the reality has never quite lived up to the fantasy. Networks did achieve limited success in all these tasks and many more, but they're still confined to a few specialized uses. They've been particularly successful in making sense of the geophysics connected with oil exploration, I believe, but we all know that machine translations of language are often terrible. Every now and again claims are made that they're better than the experts at diagnosing the cause of abdominal pain, for instance, or picking up the warning signs of child abuse, but it is mostly still very much a matter of 'research in progress.' And few physically real networks have ever been built. People hardly ever connect up working models of nerve cells, for instance. What they mostly do is program a digital computer to simulate what an artificial network would be expected to do if it had been constructed. So most research into the role of neural networks in brains is dealing, not with models of brains, but with models of models.

Complications were introduced into these digital models — extra layer(s) between input and output, more elaborate interconnection, and so forth. Some of these produced gains, but nothing that dramatically improved the utility of nets. Putting in one extra layer seems to have been the most effective innovation. What the complications did do was to make it far more difficult to predict or analyse net performance; so people began to look at it in terms of attractors. The most complex of artificial networks, however, has nothing on the complexity of those in real brains. Real neurons can have thousands of interconnections, whose strength and other properties are affected by tens, if not hundreds of factors. Moreover individual neurons can themselves perform their own personal computations (they possess some fourteen independent or semi-independent mechanisms which could allow them to do this, though it is not known how many of these are actually important), which in turn are liable to affect how they influence other neurons. If the behaviour of artificial networks can often be described only in terms of the attractors they are found to contain, then this is likely to be even more true of networks involving living nerve cells.

What people know about artificial networks naturally colours their view of ones in the brain. So it is supposed that long term memories are encoded in patterns of inter-connection and/or in the strengths of connections between neurons.[1] Similarly, attractors in real brains are considered equivalent, if not to memories viewed as static repositories, at least to memories in action — being stored and recalled. It's widely speculated that they are involved in all sorts of pattern recognition and perception, which inevitably involves previous experience and learning (that is, types of memory), just as Freeman extrapolated from his rabbits. There's quite a cottage industry devoted to modelling the attractor dynamics that may exist in a part of the brain called the CA3 region of the hippocampus. This area proved especially attractive to modellers because it is already known to be involved in memory, while its structure has been studied more intensively than that of most other parts of the brain. The sorts of attractor that modellers like to deal with are point-like ones, or the simpler varieties of periodic attractor. Strange attractors are a bit of a nightmare because no-one knows quite what they might get up to in computational terms.

Modellers are thus caught in a bit of a double bind. If the brain does behave chaotically at times, it is likely to harbour strange attractors which must inevitably make its behaviour more or less impossible to model in any straightforward way. On the other hand, if it isn't chaotic, they lose the enticing vistas that accompany deterministic chaos, especially those that go along with fractal structure. Such structures allow all sorts of potentialities for encoding and retrieving information. These include the possibility that some form of holography might be at the basis of memory, a prospect that has a lot of appeal to a small group of theorists who include Karl Pribram, one of the all-time neuroscientific 'greats.' But it is not only the prospects for understanding memory that appeal. The *doyen* of modern study of the neuropsychology of emotion, Jaak Panksepp, wrote somewhat wistfully from his American base: '... the neurodynamics of the various basic emotional states may *eventually* [my italics] be visualized as topographically unique chaotic attractors ...'

Of course it would all be much easier were it possible to look at real attractors in real brains and see what they do. However, attractors are not 'real' in any easily accessible sense. They're like laws, but laws that are both confined to particular systems and are liable to change into different laws as the system evolves. So what's required is to discover patterns of lawful behaviour. When eighteenth century botanists wanted to find out about plants, they could go out and pick them, compare different ones, and see what common features or patterns they could come up with. People wanting to study attractors, however, have to search for their ephemeral appearance in rapidly evolving systems and try to relate this to function. They are seeking to understand, not patterns as such, but patterns of patterns, which is a much more difficult undertaking. A modeller's lot is certainly challenging and can be fun, but is not altogether a bed of roses! Perhaps paradoxically, the best way forward does not lie in trying to reduce everything to its simplest form. It is to be found in going up yet another step of the complication ladder and thinking about patterns formed by the patterns of patterns — in other words, thinking about attractor landscapes.

Small, simple nets can harbour several point attractors, each equivalent to a particular 'memory' held by the net. In fact there's a good deal of theoretical work which allows people to specify the maximum number of memories that can be held by a given net, provided they know its exact size and connectivity. This is equivalent to being able to specify the maximum complexity of the net's attractor landscape. Other features of the landscape, such as the relative heights of its hills and valleys are more difficult to predict from first principles. To discover what they are like, it's often necessary to simulate the network on a computer and see how it behaves. Of course brains are far too complex to allow any hope of predicting details from first principles, while simulations have only very limited applicability. In any case, it is usually uncertain whether a supposed simulation is in fact anything like the 'real' landscape in the brain. So why, given this background, might it be useful to think about landscapes in the brain? You could be forgiven for supposing, from what I've said so far, that doing so could never be more than cloud castle building.

The best answer to the question that I know of can be found in the work of Peter Henningsen, a Canadian mathematician.[2] Building on the usual assumption that attractors are equivalent to memories and organize perception, he points out that the attractors themselves will form their own, higher level, network which will have its own dynamic. Those at the basic level result from the dynamics of networks of individual neurons. Go up a step and these basic-level attractors can be regarded as networks of neuron-equivalents, which will interact resulting in the emergence of higher level attractors. What this means is that bottom level attractors — let's think of them as individual memories or percepts — will self-organize into what will appear to be goal-directed behaviour, the 'goal' in question being the higher level attractor. Meanwhile, individual low level ones will appear to switch on and off, or fluctuate in influence, in accord with whatever the higher level dynamics may be. Henningsen's nice term for this behaviour is 'the dance of the flip waves.'

We've already come a long way, because this picture shows how lower level activities can *self*-organize into higher level ones. And this is something that is extremely hard to explain in terms of more traditional views of how the brain works. Tradition often has it, for instance, that hierarchies of 'reverberating circuits' do the organizing. People liked to produce those clunky flow charts with lots of boxes having labels like 'input,' 'executive control,' 'comparator' and so forth. They were often objects of great pride to their creators, the more complicated the better in their view. But, to most of us, there never seemed to be any satisfactory way of getting from accounts like that to the fluid suppleness of our personal experience. Attractor dynamics provides a far more natural seeming picture. And it does away with (nearly) all the cumbersome boxes, since 'control,' 'comparison' and the like are emergent properties of everything else that's going on. The whole system functions as an integrated unit, honed by evolution to work in a way that fits us for the world. This is a big step forward, and there's more to come.

Although inputs to this system, such as those from eyes or ears, will continually be changing and thus triggering the activation of existing Level 1 attractors (or occasionally the formation of new ones), the pattern of Level 2 attractor activity either may, or may not, itself alter in response to these. So the brain's activities may appear to be dominated either by inputs or by intrinsic factors. Thus, while I'm concentrating hard on writing this, I may appear to not notice the doorbell ringing. But a 'that was a doorbell ring' Level 1 attractor is nevertheless activated in my brain and I may well have a vague feeling later on that something happened, a doorbell rang perhaps, while I was typing. At the time, it didn't alter my overall dynamics, though it probably would have done so if other attractor/memories like 'I'm expecting an important letter' had been active and able to interact with it.

Henningsen does still need two clunky boxes in his model, their labels 'attentive function' and 'destabilizing function.' It's likely that he needs them because he doesn't look beyond a Level 2 dynamics, or did not do so at the time I wrote this (his ideas are evolving and the most recent version is generally posted on his website: www.netofans.net). In fact there's probably a 'Level 3,' which may well encompass the two labelled functions and prove just as self-organizing and 'organic' as Level 2, but maybe it's best just to take one level at a time when developing ideas like these.[3]

Anyhow, whatever its neural basis, the attentive function is needed to direct the flow of the Level 2 dynamics so that one can decide whether to type or listen for doorbells, for instance. It simply corresponds to what we ordinarily mean by attention. The destabilizing function is necessary because attractors at both levels 1 and 2 will often be liable to get stuck, so to speak, preventing new material from being dealt with by the brain. So, for attention to work, there has to be a means of de-activating unneeded attractors. There has to be some way of shaking up the landscape, allowing it to flow into whatever new form is needed at a given time. Indeed, it's a characteristic of attractor landscapes that the more they are used, the more stable they tend to become. It's as if use wears wheel ruts into the landscape, making it more difficult for the brain to take new directions. Or, to change the metaphor, it is as though use tends to solidify everything and make it less supple.

In outline, that's Henningsen's model. What does it give us other than a more natural-seeming way of looking at how the brain might work? It seems to point in three particularly interesting directions. One is to do with what's called 'global workspace theory,' the next with the possible functions of sleep and the last allows a fresh look at some psychiatric disorders. We'll take a brief look at each of them, in that order.

Global workspace theory

Bernard Baars, a psychologist based in California (at Berkeley), is the most prominent advocate of this. He has written a great deal about it, possibly his best known book being *In the Theatre of Consciousness,* published in 1997. The title, as well as being descriptive, may have been chosen as a dig at another well known author, philosopher Daniel

Dennett, who was saying at the time that consciousness is really nothing like a play staged in a theatre. The basic idea that Baars advocates is very simple, namely that there are lots of brain areas or modules busy processing information of all sorts in their own little backyards, and also competing for access to a centralized distribution unit. If a module gets to win the competition, the information it contains is broadcast to all the other modules. The flow of information that gets to be broadcast in this way is the flow of consciousness, said Baars. Dennett himself now proposes that consciousness is likely to be what he calls the flow of 'fame in the brain,' which is much the same idea as the global workspace one. But he still doesn't go along with the theatre suggestion, rightly pointing out that the metaphor doesn't work because the action is its own stage and audience.

It's an intuitive sort of picture, and there's a good deal of technical evidence from a wide range of sources, brain scans, psychological testing, perceptual illusions and more, to suggest that there's something basically right about it. It's steadily gaining popularity at present as people look to it to provide an adequate basis on which to build a proper theory of consciousness. A good theory won't be coming along tomorrow, most people suppose, but when we do get one it is likely to incorporate a global workspace in some form or other.

Being psychologists of a relatively mature generation, both Baars and many colleagues interested in his ideas were over-fond of clunky flow charts with many labelled boxes. And therein lay a problem. For it was never at all clear how the winning modules could achieve their success, nor exactly what 'success' involved in neural terms, nor how all the hierarchies of boxes could co-ordinate their activities.

But there's no problem if one looks at it in terms of Level 2 landscapes. The landscape is a global workspace emerging from the activities of Level 1 attractors, which can be thought of as equivalent to Baars' local modules. Those Level 1 attractors that get to have most influence on the Level 2 landscape at any given moment will automatically have the information that they contain taken up into the landscape and thus able to influence all other Level 1 attractors. It looks as though the whole of global workspace theory is implicit in the attractor dynamics picture in rather the same way that Newtonian gravitation is implicit in general relativity. Attractor dynamics explains why a global workspace appears to exist, and why it shows the overall behaviour that it does. I think that one very good reason for taking attractors to be more than cloud castles. But it's only the first of many.

The dynamics of sleep

I need to give a bit of background on this before getting to its relevance to attractors. As is well known, there are two different varieties of sleep, slow wave and rapid eye movement (REM), which have a whole lot of very different characteristics. One difference is that vivid dreams occur during REM, but any dreams associated with slow wave sleep are more like gentle cogitation than action-packed adventure. No one knows why either variety occurs. Ideas that it's all about resting the whole brain have generally been abandoned, partly because some bits of the brain tend to be more active during REM sleep than when awake. But maybe slow wave sleep has something to do with cortical rest and REM sleep with brain stem rest, because reduced activity in cortex and brain stem is separately associated with these two varieties. That's so vague as to be not much help, though.

One thing that is clear is that sleep has something to do with memory and learning, particularly learning skills (that is, 'procedural' memory). There's lots of evidence provided by psychologists, showing that if a person or an animal learns some skill, subsequent testing will show him/it to be faster and more accurate at the skill after a period of sleep than after the same amount of time spent awake. Indeed performance the next morning, or whenever, can often show improvement over that measured immediately after learning the skill. The same seems to be true, to a somewhat lesser degree, of memories for things like facts and figures (that is, 'declarative' memory). This evidence is usually interpreted to mean that sleep helps memory 'consolidation' — the fixation of short-term memory into long-term format.

Sleep deprivation experiments have thrown less light on the functions of sleep than some had hoped. People kept awake usually feel rotten and become less accurate and more stereotyped in their performance. At one time it was said that REM sleep was the *really* important sort that caused all the detriment if you didn't get it. Certainly it seems to be true that, if sleep deprived and then allowed to rest, you will catch up on missed REM sleep before catching up on slow wave. But more recent work seems to be showing that missing slow wave sleep can be at least as bad for you as missing REM. The whole picture is clouded by the fact that some antidepressant medications can completely suppress REM sleep (or at least they can suppress all evidence of REM sleep — just maybe whatever REM does is still continuing somehow, despite absence of evidence for it.). Suppression can continue for weeks or months on end without causing obvious ill effects; indeed the antidepressant effect of these drugs is weakly correlated with the suppression of REM, while simply depriving people of sleep, especially REM sleep, without giving any medication can sometimes temporarily alleviate depression.

Most people have probably read popular accounts reporting that, if you completely deprive rats of sleep, they often die after a few days. These experiments were wholly indefensible and disgusting, if only because whether the deaths were due to sleep deprivation as such, or to stress caused by the methods used to keep the animals awake, is not clear. It is very hard indeed to understand how the people who conducted these experiments could ever have hoped to get clear answers to their questions. It should perhaps be said in the experimenters' defence that most of their work was done in a cold war context of concern over the use of sleep deprivation in 'brainwashing' prisoners. But that's still a pretty poor excuse for torturing animals to death, especially when the information obtained could easily have been foreseen to be valueless.

Even though no one knows for sure what it does (other than possessing a role of some sort in memory 'consolidation'), sleep must have some absolutely essential function since all higher animals do sleep, despite the huge cost to some of them in Darwinian terms. Questions of survival modify sleep patterns but, whatever the risks involved in sleeping, no animal has been able to dispense with it. Big carnivores like lions, who are not at risk, tend to sleep a lot maybe in order to conserve energy. Animals like us, perhaps safer tucked up in a tree at night than roaming around on the ground, sleep less than lions and usually at more regular times. Animals such as antelopes, as much at risk from hyenas during the night as from lions during the day, tend to sleep in short snatches, standing on their feet while herd mates are awake. The most extreme adaptation is shown by dolphins who would drown if they went soundly to sleep like us. They are said to sleep one hemisphere of their brains at a time, leaving the other awake and able to swim and watch out for killer whales. Moreover their infants, even more at risk than adults, sleep less than adults, which is the reverse of the pattern in all land mammals.

That's the background. Sleep does have some role in improving learning and memory but it's not clear what the role is, nor why sleep is so essential. No one knows why there should be two such very different forms of sleep. The evidence about whether they have different roles, and what the roles might be, is fuzzy at best. Time to bring on attractor dynamics, which can explain why sleep is necessary and what it achieves in terms of improving learning. It also hints at why there should be two different forms.

As mentioned earlier, it's an inevitable feature of the attractor landscape that 'grooves' worn into it tend to become ever deeper, thus making it more difficult for the 'destabilizing' and 'attentive' functions to do their bit. As ruts multiply and deepen, these functions will tend to be ever more driven by the lower level dynamics. And, of course, this process will occur every day in every animal, as familiar environmental inputs trigger well practised Level 1 dynamics, which in turn give rise to habitual Level 2 dynamics. If the process were allowed to continue indefinitely, an animal's brain would probably end up dominated by one huge, unchanging attractor, manifesting in a fixed and rather simple landscape. The animal would lose all flexibility of behaviour and would quickly succumb. Very tempting to suppose, therefore, that sleep is what saves us from this outcome. After all, peoples' behaviour does become more stereotyped when they are sleep deprived, which is surely an early sign that the grooves in their landscapes are deepening.

Suppose the main function of sleep is indeed to freshen up attractor landscapes, what would it need to do? One essential would be to shake up Level 1 landscapes, so as to balance them and remove any pits and potholes. Another would be to 'tune' Level 2 dynamics so that they could deal appropriately with any new features that had appeared in Level 1 during the day. The function of sleep according to this picture is thus not memory 'consolidation,' in the sense of 'fixation,' but removal of any excessive fixation plus help with getting new memories to fit in appropriately with Henningsen's 'dance of the flip waves.' These actions result in the improvements in task performance and flexible recall seen after a period of sleep. The apparent improved 'consolidation' of new memories on waking, found by psychological testing, is in fact a consequence of diminished interference by other memories in Level 1 landscapes and improved capacity and accuracy of Level 2 landscapes to orchestrate the recall of new memories when required — or so the attractor dynamic picture suggests.

It's also tempting to go a bit further and suggest that slow wave sleep is what smoothes, soothes and balances Level 1 landscapes, while REM sleep is the 'tuning' process.[4] After all, people have long suspected that dreams may have something to do with 'rehearsing' memories or neural programs of some sort — which is simply a different way of putting the 'tuning' suggestion arrived at here. Going a step further still, one can speculate that something is wrong with the Level 2 dynamics in (some) people who are depressed. Thus, if they are prevented from 'tuning' new material into their abnormal dynamics (by depriving them of REM sleep), their depression may temporarily be alleviated. But, coming back to earth, is there hard evidence for any of this?

One very basic and straightforward prediction to be made from this picture is that sleep deprivation should result in reduced EEG complexity, because deprivation supposedly allows the attractor landscape to evolve into a simpler, more 'stereotyped' state which is likely to be reflected in the EEG. That seemed the sort of thing someone, somewhere, might already have researched at some time; so it was back to Google. There were lots of vaguely relevant research reports, but the specific information wanted didn't seem to have attracted a great deal of attention. Nevertheless one study, carried out in 2001 by a group based at Yale (see Jeong et al., 2001, under Further Reading), seems to have shown precisely this predicted complexity reduction in subjects whose EEGs were examined after one night's sleep deprivation, and compared with others who had not been so deprived. While it is true that one swallow does not make a summer, at least its arrival may show that summer's on the way. Looking at the possible dynamics of attractor dynamics has allowed a successful prediction about real brains. That's surely an indication that the field may be moving on from the eighteenth century towards a nineteenth century stage of science!

Psychiatric disorders

There's a problem to be faced when writing on this topic. It does not readily blend with attractor dynamics, which is about how the brain really works. A lot of psychiatry, on the other hand, is about superficial appearances when something unknown goes wrong. I'm going to try to describe how the dynamics might relate to a few particular psychiatric disorders, and this is especially problematic because their accepted classification is ... I was going to write 'crap,' but maybe that's a bit harsh. A lot of effort has gone into trying to delineate sensible categories. It's just a pity that we have made only small advances from the position reached by Emil Kraepelin (the 'father' of psychiatric classification) a hundred years ago. Things were actually slightly better in some ways thirty or forty years ago than they are now. At that time lot of research effort was going into identifying meaningful categories of illness and a whole range of alternative classification systems were under test.

Nowadays, ideas have tended to rigidify and meaningless tautology has crept in. It is often implicitly assumed, for instance, that 'depression' is what a patient has if he/she is considered to need antidepressant drugs, or that 'neurosis' is what Medicare agencies can be persuaded to fund the treatment of.

Of course, great hopes are pinned on identifying meaningful categories of mental disorder through genetic studies, research using brain scans, and so forth. So far, this has not paid off. You still can't go for a blood test or a scan which will tell you whether you've got 'real' depression or are merely having a bad week and will be better soon. Maybe pay-offs from these lines of research are imminent and the whole classification will get sorted out, but I'm not holding my breath. Some advances can probably be expected. On the other hand, the history of psychiatry over the last forty years is littered with similar high hopes that came to nothing. My guess is that progress will be a matter of small steps only.

It's worth taking a brief look at some of the reasons for this situation. Obviously the main one is the sheer difficulty of psychiatry compared to many other branches of medicine. It sits at the interface between so many disciplines — internal medicine, neurology, psychology, sociology, to name a few — that it would be surprising only if it *was* clear-cut. A more specific impediment to progress has to do with the fact that nearly all the conditions that psychiatry treats are more analogous to 'headache' than to a specific disease such as 'appendicitis.' Headaches have a thousand causes, indeed are often quite normal parts of human experience, and so it is with so-called 'illnesses' like anxiety state or depression. Even conditions like 'schizophrenia' are probably more akin to 'acute abdominal pain,' say, than to 'appendicitis'; acute abdomens can have tens of causes, of which appendicitis is just one.

Another reason for the problems provides a quite instructive example of how good intentions can go wrong. The most influential psychiatric diagnostic system is the American 'DSM.' There's also an 'International' one, which has tended to get ever more like DSM with each successive revision. Back in the 1960s, the compilers of DSM3 were faced with various problems, including the difficulty of distinguishing 'illnesses' from the everyday vicissitudes of life. They adopted several sensible measures for coping with this, but they also faced another big problem. It had become apparent that psychiatric diagnoses were very unreliable — two psychiatrists faced with the same patient were quite likely to give two different diagnoses; so-called 'schizophrenia' in New York included nearly all the group of patients who, had they lived in London, would have been diagnosed as 'manic-depressive.' The DSM3 compilers therefore deliberately opted for a diagnostic system sufficiently precise and rigid to give good reliability of diagnosis — so two different psychiatrists, using the system, *would* agree.

The compilers were well aware that their illness categories had no more than surface validity; they were more like 'a bad headache, lasting three days' than like 'pain in the head due to temporal arteritis.' They knew that the psychiatry of their day did not allow identification of truly valid illness categories but they hoped that, if psychiatrists could at least agree on classifying patients in some way that appeared reasonable, this would facilitate identifying 'real' conditions in the course of time. Maybe their hope was reasonable, but events overtook them. These included the rise and rise of both the psycho-pharmaceutical industry and hospital and insurance bureaucracies; all agencies which demanded rigid 'diagnoses' to feed their systems. Hence, what was acknowledged to be a stopgap system by its originators has become virtually set in stone. There have been revisions to DSM since DSM3, but they have only tinkered with it, mainly because no-one yet knows how to produce anything fundamentally better. There is little doubt, though, that the rigidity of the system actually impeded, and still impedes, the search for valid categories, which is entirely the opposite of the original intention.

Because the meaning behind psychiatric diagnoses is so shallow, it's not possible to point to specific conditions and say 'this may be due to that feature of attractor dynamics.' You can only take an overview of what may be going on. Many mental disorders are characterized by a narrowed range of mental activity, within which particular features become abnormally frequent or intense. For example, in depression there's a general bias towards the gloomy side of experience, often accompanied by prolonged rumination over particular misfortunes or fears. Anxiety neurosis is characterized by bouts of overwhelming worry and fear, sometimes coming out of the blue, sometimes triggered by specific factors (phobias). In obsessive-compulsive disorder, repetitive ruminations or compulsive behaviours get in the way of everyday activities and cause great distress.

Thought of in terms of the dynamics, all these conditions would seem to have distorted 'landscapes,' in which particular attractor basins have grown too large or too deep. Why should this have come about? It might sometimes result from normal processes of memory and learning, when what is learned is adverse. Trauma might produce it, and so might milder but more repetitive occurrences, such as repeated failures or frequent minor scares. Apart from environmental causes like these, the distortions could also result from failures of the 'destabilizing' system that normally promotes landscape flexibility, or of the 'tuning' system which helps to integrate new attractors into existing landscapes.

Alert readers may notice that the suggestion about tuning system failure appears to clash with my earlier speculation that REM sleep deprivation may improve depression by interfering with 'tuning.' However, tuning presumably works to accommodate new material to any existing landscape. If a landscape is already abnormal, it will tend to fit new stuff to conform with abnormality. So interfering with the process in people already abnormal may help their symptoms. On the other hand, substandard tuning in people with hitherto normal landscapes may make it easier for abnormalities to grow.

I suspect that destabilization system failures or weaknesses may be especially important in predisposing to many of the anxiety/obsessive types of psychiatric illness, and might often have some biochemical basis. But this is speculation only. It merely invites that well-worn comment: 'more research is needed.' Attractor landscape theory may have more immediate relevance to another large class of psychiatric abnormalities. These are the so-called 'dissociation disorders.' What can be pictured as happening in them is that chunks of landscape get functionally disconnected from the rest. Perhaps the most dramatic example of these conditions is multiple personality disorder, in which different people appear to inhabit the same body at different times. Scary! Typically, some of the sub-personalities are aware of their fellows while others are not. The record for the number of sub-personalities housed in one body keeps going up. It started at three with *The Three Faces of Eve* (the best-selling, 1957 book by psychiatrists Corbett Thigpen and Hervey Cleckley, which appeared to set the whole ball rolling), but now stands at over twenty I believe. There have been claims for even more — up to sixty or so — but it's hard to see how these could possibly have been authenticated. The condition is thought to be often based on childhood trauma but is also clearly a cultural construct, which must be 'learned' and rehearsed, since it is rare outside North America.

What's needed in these conditions is to re-establish connections. In the case of multiple personalities this would involve reconnection between the different sub-personalities, which is already the goal of standard therapies for it. Peter Henningsen, however, has suggested that a non-standard treatment might prove especially effective. The aim here would be to wear a 'valley' in the overall attractor landscape, common to all personalities, which would thus connect (or re-connect) them. He suggests that a way to achieve this might be to train each personality with some repetitive mantra, in the hope that each would wear the same groove in the overall landscape, thus giving the separate personalities a 'channel' through which to communicate. Worth a try maybe, though the assumption that a single mantra practised by separate personalities would wear one common valley, rather than separate valleys within the landscapes of separate personalities, could prove over-optimistic.

Whether or not this particular suggestion does prove useful, it is at least clear that the new perspectives provided by attractor dynamics could turn out to have very practical, down-to-earth applications. Those 'castles in the clouds' are looking ever more solid after all. In the next chapter, we shall go a step further and suggest that they really are at the basis of our experience in a quite fundamental sense — and nothing is more real to each of us than the flow of our conscious experience.

4. Death and Other Experiences

Joe Smith was having a bad day. It had started off all right; a crisp, bright, frosty morning and he had gone out to scrape the car windows before going to work. Last night's indigestion seemed to have gone. 'Must have been that second helping of pizza,' he thought to himself. He was just reaching across to put the finishing touches to the windscreen when something really bad happened. A crushing chest pain that made him gasp. He could feel himself stagger and saw that he had dropped the scraper. Then everything went black ...

No, he didn't die. There were hazy memories; an ambulance siren; Emma's face as she bent over him; being rushed through doorways on a trolley. Everything was so jumbled that he could not keep track of what was happening. The pain had receded a bit — it was more of an ache now. People were bending over him, doing things and talking to one another, though he could not seem to follow what they were saying.

Then, quite suddenly, everything became much clearer. 'Looks like a large anterior infarct,' said a man's voice quite close to his ear. 'We'll have to keep a close eye on him. Here, hand me that streptokinase, will you.' Joe liked the voice, it sounded calm and competent. And the pain had completely gone, which was great. He was really feeling very well indeed, except ... Well, except that he seemed to be floating about four feet above his body. He could see it stretched out on the trolley, with drips going into the left arm and cardiograph leads attached to the chest. The man with the calm voice was standing at the body's side. Though dressed like a surgeon, he wasn't wearing a cap and had quite a large bald patch, Joe noticed. But he was feeling so pleasantly calm and content himself, that he couldn't be bothered to pay much attention to the medical drama. His gaze wandered up towards the ceiling. There was a dark patch up there, getting steadily bigger. It seemed to reach out and engulf Joe before he could do anything about it, and he felt himself being carried along at great speed through some sort of tunnel. Then a patch of light appeared ahead of him, which grew steadily brighter; so bright that it ought to have hurt his eyes, but didn't.

The next moment, he found himself standing in what seemed to be warm sunshine. When he looked about, he was surrounded by the most beautiful parkland he had ever seen. Feelings of love and joy swelled in his heart. There were people coming towards him. Surely that was Mother, dead these last nine years, and there was Emma, and that older man looked like his Grandfather. He yearned towards them, but something held him back. Suddenly he seemed to be falling, falling back into his body. Clunk! He was there. It was an unpleasant sensation, like putting on clammy, muddy clothes. His chest was aching, and everything became confused and hard to follow once more. Nevertheless he did gradually get back to normal and, years later, could still remember with the utmost clarity what 'death' had been like. But he could never recall much about the rest of his time in hospital or his convalescence. On balance it had probably not been such a bad day after all, he considered, for it seemed to have taught him a great deal about what is important in life and what is not.

Explaining Joe's experience

Joe's account is of course fictional, but is typical of a great many stories that people have really told about their 'near death experience' (NDE), as it is called. His phenomenology was quite varied, as people do not always get all the features that he experienced; many get only one or two items out of the range of sudden calmness and clarity, floating, tunnel travel, 'seeing the light' and meeting relatives in the suburbs of paradise. On the other hand, some accounts are richer than Joe's. Many people have met religious figures, perhaps Christ or Mary if they are Christians, maybe Krishna if Hindu, and some hold conversations with people they encounter. A few have undergone a 'life review' in which they have re-experienced, seemingly in a very short time but with great intensity, many of the good and bad episodes or actions in their past lives. An even smaller proportion have claimed to experience bad actions from the perspective of their victims.

Jung had already got in on the act long before the great surge of interest in NDEs, which was triggered by Raymond Moody's 1975 book *(Life after Life).* This reported on about 150 cases that had come to Moody's attention in the course of his work. He was a philosophy lecturer who subsequently became a psychiatrist, and was initially told by some of his students about NDEs they had had. He later set out to collect more cases, and was surprised by how frequent they seemed to be. But to get back to Jung, he suffered his heart attack in 1944, thirty years before Moody's book came out, and subsequently described a particularly rich

NDE in the course of which he saw the earth from orbit. He later calculated his 'altitude' from recollections of his field of vision at the time. It turned out to have been around 1000 miles! He then encountered a huge meteorite, carved like a Hindu temple, which contained, Jung felt, the answers to all life's most important questions. He was about to enter the temple when a vision of his doctor, who had taken the form of 'a *basileus* of Kos' (a king-like figure belonging to the alleged birth-place of Hippocrates, the 'archetypal' physician), summoned him back to earth. Maybe his experiences were atypical because there were elements of delirium in them, missing from more usual NDEs. He continued to get night-time visions of beautiful scenery and numinous people for quite a while after the NDE itself, symptoms that strongly suggest a toxic psychosis of some sort.

Not everyone who has a close brush with death has an NDE; the proportion in coronary care units doing so seems to be somewhere around 10% — maybe up to 20% according to some reports. Moreover people report seemingly identical experiences as having occurred when their lives were not at risk. These may actually outnumber the brush-with-death ones according to the Religious Experiences Research Centre, an organization formerly based in Oxford but now at the University of Wales, which has an archive of over six thousand case reports. However I want to concentrate on the 'traditional' NDEers for the time being, because they make better examples when it comes to trying to explain their experiences. They don't have quite so many 'degrees of freedom' as the others, to use a bit of jargon from Chapter 2, so it's not quite so difficult to tell what may have been going on.

One thing that does seem clear is that NDEs are not creations of modern resuscitation techniques or other features of Western culture, since they have been reported from a wide range of both contemporary and historical sources. They seem to be a potentiality built into human nature, sometimes triggered by life threatening illness or trauma but able to occur in other circumstances, too. This generality is very important because it cuts out a whole range of potential cultural and social explanations.

Unsurprisingly, the content of the experiences *is* influenced by culture. A Zambian woman is said to have entered 'a calabash' instead of a tunnel, for example. The religious figures whom you are likely to meet in 'paradise,' should you arrive there, seem to be at least partly determined by your background, while relatives and friends are of course specific to personal history — nor do they have to be themselves dead in order for you to meet them. There's been a suggestion that the NDEs of Westerners are more likely to be action-packed than those of people from other cultures. Nevertheless, whatever the influence of culture and personal history on details, there seems little doubt that the ability to experience NDEs at all is independent of culture. What *can* be made of them?

Religious believers often think they have no problem explaining what happened. Their souls were hoovered up heavenwards, they suppose, and got a glimpse of the afterlife. Some of those lucky enough to attain the more usual, paradisal experience were so impressed by it that they converted to religious belief, which previously they had lacked. Consistent with this interpretation, so traditional or fundamentalist believers might claim, it is sadly true that a very small proportion of NDEers end up in distressing surroundings. Some are simply upset or frightened by the tunnels and lights but a few, probably no more than 1–2%, find themselves in hellish surroundings, complete with screaming souls and grinning skulls according to one account.

Although it fits the reported facts and is consistent with the preconceptions of many people, there is a glaring problem with any explanation along these lines. How could a soul so far from its body, allegedly around 1,000 miles in Jung's case, impress such vivid memories into the brain of that body? After all, it is through neural memories and mechanisms that people later report their experiences. And people don't forget NDEs, it seems, whereas memories of the rest of their illnesses, when their souls were presumably still housed in their bodies and thus in a better position to interact with them, are usually scanty and quickly fade. This is perhaps the main inconsistency in soul-based explanations of NDE, though there are many other more technical objections; especially that the explanation would apparently violate energy conservation.

Many scientific explanations, on the other hand, seem unconvincing or incomplete for what amounts to much the same principal reason. They don't violate energy conservation, of course, but they do have problems accounting for the clarity and memorability of NDEs. There's a long list of attempts to explain. I won't go into them all, but it is worth mentioning the front runners.

NDEs have something to do with disturbance or convulsive activity in the temporal lobe of the brain, it has been suggested, maybe triggered by oxygen shortage associated with the life threatening event. This has some plausibility, since people with temporal lobe epilepsy are prone to religious preoccupations and even ecstasies, while electrically stimulating parts of the temporal lobe can sometimes result in 'mystical' experience. Others have claimed that the feelings of peace and love accompanying many NDEs are signs of endorphin (the brain's own opiate) release in response to trauma and, in such extreme circumstances, perhaps enough endorphins get produced to cause the other experiences, maybe with a little help from 'shock' and anoxia. This, too, is plausible as far as it goes. Susan Blackmore has developed an especially elaborate model along these lines, involving oxygen shortage and the dying brain's attempts to make sense of what is going on in the best way that it can.

Another suggestion, which I think particularly interesting, is that NDEs are a sort of acid trip caused, not by the LSD that used to be (still is in places!) such a feature of the club scene, but by an equally powerful hallucinogen named DMT (dimethyltryptamine). It is related to LSD chemically as well as in producing similar effects. We all of us generate small amounts of DMT all of the time, as a breakdown product of one of the neurotransmitter chemicals in our brains (serotonin). But it does not normally have any noticeable effect on us because it is very quickly destroyed by special enzymes (MAOs, monoamine oxidases). These enzymes exist in the brain and get rid of any DMT produced locally there, and are also found in the gut and the liver. The latter destroy DMT that we eat, since there are small quantities in quite a range of dietary items, especially some plants. So the idea is that we sometimes produce too much DMT for the MAO enzymes to cope — and then we're off on our own, personalized trip. Maybe there is something about being on the verge of death that makes us specially likely to generate excess DMT, is the final step in the argument.

It's quite likely there is a bit of truth behind all these explanations. Each is not implausible and fits some of the facts. Some parts of some NDEs are a bit like the 'mystical' experience of temporal lobe stimulation. Oxygen lack, pain and shock can trigger all sorts of biochemical events, including endorphin release which could certainly be responsible for some of the feelings associated with NDEs. DMT might well account for the hallucinatory aspects if it is indeed over-produced, which is a possibility only — no one knows for sure. But there is something missing. People who are shocked, anoxic and in pain are normally confused, their consciousness is clouded and, above all, they usually have very poor recall of their condition. Yet NDEers claim great, often super-normal, clarity of consciousness during the experience and, consistent with this, their recall is excellent.

To account for the discrepancy, it has sometimes been suggested that people concoct the experience later and that NDEs are a type of false memory syndrome. Maybe that's just about believable of people who happened to have read, and been impressed by, Raymond Moody's book for example. But we've already seen that NDEs are not culture dependent, except as far as the 'small print' of their content is concerned. So this particular 'explanation' doesn't hold water.

What about an alternative? LSD experiences are often very intense and memorable, and maybe DMT experiences are the same. DMT can and has been given to normal people, either intravenously in a sufficient quantity to get past brain enzymes or alternatively by mouth and accompanied by something which blocks the MAO enzymes so that it is not destroyed. It does indeed produce very memorable experiences. In fact the first experience is usually to make perfectly fit people feel so nauseous and otherwise terrible that they fear they are going to die. The hallucinations, which arrive later, are often of jungle scenes, snakes and other animals. There are sometimes commonalities and resemblances between Near Death and DMT experiences, but they also differ in some respects. Moreover, although it is difficult to make comparisons, a typical NDE seems on the whole to be at least as memorable as a typical DMT experience. We're back to the problem of why this should be so, given that NDEers were traumatized, often severely so, before their experience, while people taking DMT were fit and well, with their memory mechanisms presumably in tip-top condition. It's therefore probably not the case, I think, that NDEs are primarily due to DMT overproduction. As we shall see, it is possible that both types of experience share similarities because they have common origins, not because the DMT directly causes the NDE. After all, no one has much of an inkling as to how a simple chemical, or stimulation of some brain area for that matter, could cause experience of any sort, let alone something so complex as an NDE or an acid trip.

The position we've reached, then, is that there are a number of plausible-looking candidates for triggering NDEs, any or all of which may actually be involved in doing so. We simply don't know. However, NDEs themselves are very unusual, complex and apparently highly organized experiences, which suddenly occur in settings where only haziness and confusion might be expected. What does the situation look like from the point of view of attractor dynamics?

The short answer is that it looks pretty weird, for it seems that attractors might be fully in control of the neurology. Despite the earlier suggestion about them being law-like (Chapter 2), we have hitherto pictured them as emergent properties of neural nets in the brain, with an implicit assumption that they should not be regarded as anything more than appearances controlled by the underlying behaviour of neurons. Given the idea (Chapter 3) that a 'Level 2' attractor dynamics exists, deriving from the interactions of 'Level 1' attractors, the assumption was always over-simplistic. Now it seems attractors must be taken as seriously as any other emergent property — as seriously, for instance, as the liquidity of water, which emerges from interactions between molecules of H_2O but nevertheless guides and 'controls' aspects of the behaviour of those molecules. After all, *something* must be responsible for all the delicate tuning needed to organize a traumatized brain into producing the experience of a lifetime. The only realistic candidates are attractors. And the experiences themselves offer hints about the nature of these attractors.

The initial feelings of peace that often occur in NDEs, and the later ones of love and yearning, seem much the same as those all of us get from time to time, though allegedly often more intense. The attractors involved here, no doubt sited in areas of the brain responsible for emotions, don't appear unusual. But then we get to floating above one's body and then being sucked up into a tunnel. Definitely not like everyday experience! Let's take these two in turn.

The feeling of being out of one's body (OBE) is actually not that uncommon. As many as 20% of Americans have had it to varying degrees, according to surveys. It's sometimes a consequence of severe stress; soldiers in the trenches often got it for instance. Temporal lobe abnormalities can 'cause' OBEs, or they can simply happen for no obvious reason. Evidently some brain mechanism (maybe equivalent to a Level 1 attractor) exists in many or all of us, able to produce this experience. Seeing one's own body from the outside (called 'autoscopy') is less common but is not confined to NDEs. People have speculated that it may have something to do with the 'self-model' housed in all our brains coming adrift from normal bodily feelings, like those of having a full stomach, being seated in a comfortable chair, and so forth. If the 'self-model' is not pinned down to the body by these feelings, so the argument goes, it is free to drift off into other parts of the brain's representation of space, and even to appear to be seeing itself from the outside.

Incidentally, if there is anything to this argument — and it appears reasonable — the image of themselves that people see should be like their mirror image, not like themselves as others see them, since the visual image would derive mainly from composite memories of looking in mirrors. I don't know whether anyone has ever thought to ask NDEers the appropriate questions to find out.[1] Sue Blackmore, psychologist, author and expert on NDEs among much else, also has not come across any systematic information about this but thinks the answer is probably 'yes': that is, NDEers probably do see their mirror image (personal communication).

Be that as it may, what's most strange about the typical NDE experience is not the floating out of one's body as such, but the way it is often integrated with goings on in the emergency room or wherever. When put in the terms used in this book, there's some very sophisticated Level 2 attractor dynamics needed to allow this. Of course, if people can 'see' or 'hear' things that they genuinely could not have accessed from the vantage point of their bodies at any time, as they sometimes claim, the whole situation becomes stranger yet. But I don't want to go down that route because the evidence is anecdotal only. Attempts to prove that people can access information while 'out of their bodies,' by hiding messages above the roof of coronary care units for instance, have come to nothing (so far).

It looks as though an attractor that seems to be 'hard wired' into the potential dynamics of everyone's brain (the sensation of becoming detached from one's body) is activated in the early stages of an NDE and becomes integrated somehow with normal sensory dynamics; that is, input from the NDEer's current surroundings or very recent memories. Then many of them are sucked up into the 'tunnel.' Theorists have had a field day here. One popular notion is that the NDEer is 're-living' their birth experience — travelling through a dark tunnel and finally emerging into the light. Maybe so, though NDEers usually feel that their passage through the tunnel is fast. Labouring mothers would be only too happy if that were true of most real births! Moreover NDEers usually report feelings of bliss and peace at the end of the tunnel, whereas babies generally yell loudly and appear far from blissful. Nevertheless, I suppose it's not impossible that the 'tunnel' experience represents a much abbreviated and bowdlerized memory of birth from start to being finally cuddled by Mum. It's another attractor which we may all possess, though not all of us will ever experience it in action.

After the tunnel, when NDEers get to their destination, everything usually gets more personal again. People see their own relatives or the mythical figures that played an important part in their own up-bringing. It's even more personal in the case of the 'life reviewers,' who experience a sort of rapid biopic of their own histories. The relevant attractors are thus ones that people have built themselves in the course of their lives, not universals that we all share — except for the emotional components, which are universal. But the whole thing is so amazingly well *organized.* Surely neurons must be dancing to the tune of hierarchies of attractors that the NDEers house. And the attractors themselves behave as if they have quite tenuous dependence on the bodies in which they reside.

Is an explanation of the memorability of NDEs hiding in all of this? Jung, if asked to account for it, would have said something like: 'if you enter the realm of the archetypes, you must expect to be deeply affected by whatever happens there.' We've already seen (Chapter 1) that his 'archetypes' are equivalent to attractors shared by the whole of humanity. So NDEers have entered his archetypal realm by virtue of activating their 'floating' and 'tunnel' experiences. The idea of paradise seems almost universal, too, and many of them experience that also. Maybe, therefore, there is sense behind Jung's (imagined) explanation. Going back to the language of attractors, it looks as though activating one or more of the 'universal' ones that lie dormant in all of us may dramatically affect the attractor landscape, allowing cascades of further activation that would or could not have occurred in the previous landscape. Once the landscape has altered so dramatically, it is likely to bounce back only slowly towards its previous form, and may always retain some of the new features. As it turns out, if one thinks of attractor landscapes as analogous to real ones, having an NDE makes a great impression on people in an almost literal sense.

Let's move on now, and see whether the DMT experience, so similar in some ways to the NDE one, has anything more to tell us about attractors.

The DMT experience

Joe's cousin Jim was an adventurous character in all sorts of ways; an anthropologist attached to the University of Midchester, he had made a special study of Amazonian tribes. He had paddled canoes up winding creeks, with alligators plopping into the water as he passed, had been shot at with arrows by the Yoruba before they got to know him (luckily the arrows were not poisoned, or at least the one that scratched him wasn't), had removed leeches from between his toes and endured close encounters with more snakes than he cared to remember. As he told people often enough, he was actually far more at risk from illegal loggers than from any of the Indians or the wildlife; more at risk still, like anyone going to less developed parts of the world, from that everyday hazard the road traffic accident. He fully endorsed the prevailing belief in anthropological circles that one could not hope to properly understand a culture without, in so far as possible, sharing in its everyday life. So he would disappear into some distant village for months on end, cavort in a loincloth and try to cope with the complexities of one of the most difficult languages in the world. In fact, he had more success in mastering loincloths than language, and usually ended up speaking pidgin Portuguese like everyone else — and eating out of tins, as he discovered that catching animals or finding nutritious roots is more difficult than might be supposed.

In one such village lived a shaman, famed far and wide for skill as a diagnostician and healer. He was an amiable man, who looked about fifty years old but was possibly in fact younger, willing to discuss his art with Jim. 'Ayahuasca shows me the nature of people and their sicknesses,' he reported. Jim already knew quite a bit about both shamanism in general and ayahuasca in particular, since both had been a focus of academic interest for a decade or more. The word means 'vine of the (dead) spirits.' A brew, also referred to as 'ayahuasca,' is made from this vine and the leaves of another plant in quite a complicated process involving much pounding and boiling. It is used both by traditional shamans in villages and by several religious sects in the shanty towns. Users regard it as divine or semi-divine, Jim knew, although it seems no more than a powerful hallucinogen to outsiders.

As often happens, one thing led to another and a day came when Jim was allowed by his new friend to take no food, or anything other than pure water. The aim was to purify Jim so he would be able to try ayahuasca that the villagers would soon prepare. As it turned out, the preparations started late afternoon and went on for much of the night. Jim was expected to participate, but kept nodding off due to the rhythmic thumping and chanting. He was not much help. But, towards dawn, someone shook him awake and handed him a cup of the brew. Everyone was watching as he put it to his lips and tossed it down. 'Ugh!' It was even worse than he had expected — and it was described in some of the literature he had read as the most disgusting taste possible. He retched, then ran out of the hut to vomit. Of course he didn't have much to bring up, so was soon able to return inside and sit leaning against a wall. Then the coloured lights started, pretty patterns that intrigued him so much he lost track of time ... They faded after a while, but the light from the oil lamp seemed to be getting ever brighter. Everything in the hut was glowing with its own light, indeed he could see that everything was flickering and *made* of light, and was so beautiful. The incredibly elegant curve of that pot! The loveliness of old Maria over there in the corner! The huge stars visible through the doorway!

He got up and wandered out, with some idea of seeing more. The jungle seemed aglow with cool flames; a huge and gloriously coloured python was curled about an old stump; there were jaguars to be glimpsed further off, slinking among the glowing tree trunks. Then the scene changed and he seemed to be looking at some lovely, elaborate city of pinnacles and spires, all elaborately carved. Angkor in Cambodia must have been like that in its heyday, maybe. And there were brightly dressed people moving among the buildings. He sat down again, leaning against the outside wall of the hut this time, and gradually the vision faded. He slowly came back to his normal self, feeling as if he had been washed and purified, inside and out. Daylight had come two or three hours previously.

Jim made many more trips to ayahuasca land in subsequent years. For what he saw, heard and felt there seemed worth all the initial discomfort of the sessions. And, apart from the magic of the visions, he felt strengthened in his very soul by each visit. But it is time for us to leave him. We need to think about his experience and what it might mean.

The potion that produced these effects is chemically complex, but its activity is thought to derive from two main ingredients; the first DMT, which is present in the leaves added to it; the second a range of harmaline alkaloids from the ayahuasca vine itself. The main effect of these alkaloids is to block MAO enzymes that would otherwise destroy swallowed DMT before it could reach the brain, but they also have weakish hallucinogenic properties of their own. How Indians ever hit on the idea of combining the two plants to produce their remarkable brew is a mind-boggling question. After all, there are thousands of plant species in the Amazon, and the number of possible combinations of any two (and the means of processing: soaking, boiling, pounding, and so on) must run into millions. Maybe a significant proportion of combinations would prove psycho-active without also being lethal, and they just happened to hit on a particularly powerful combination — nevertheless it's very hard to imagine how they got there. Shamans themselves are said to have claimed that the 'spirit of ayahuasca' guided them to its discovery, and maybe that's as good an answer as any.

Like Joe's NDE, Jim's experience was fairly middle of the road for a first ayahuasca trip; more elaborate than some but less so than others. With practice, users apparently learn to direct the flow of their experience to a degree — hence some shamans can use it to 'gain insight' into sick people, others can make visionary journeys and so forth. It can readily cause OBEs (out of body experiences). Indeed Rick Strassman, a psychiatrist who made a clinical study of the effects of injected DMT, seems to have recruited some of his volunteer subjects, who were mainly experienced drug users, partly because they were hoping to get OBEs. They were mostly not disappointed. Autoscopy (that is, seeing one's own body after 'leaving' it), on the other hand, while common in NDEs, is rarely mentioned in accounts of the DMT experience.

So far, you are probably thinking there's not that much difference between the two types of experience. NDEs cannot be compared with practised ayahuasca sessions, which often are rather different and partly under voluntary control, because few people get to have repeated NDEs. Any repetition, and it's only too likely to be more than *near* death. So comparisons must be made with what naïve drinkers experience. There's more of a light show, usually, with ayahuasca, but tunnels and white lights don't feature so much, while paradisal scenes and experience are generally integrated into the drinker's actual surroundings, instead of taking place in another land beyond the tunnel. Moreover, there is less of a coherent 'story-line' with ayahuasca. People are plunged into marvellous scenes, but it's not often clear what they mean. Any visionary figures are likely to be strangers, not relatives, nor the Christs, Krishnas and Virgin Marys of NDEs.

It's hard to know what to make of these differences. Maybe they are an indication that NDEs are triggered by something other than DMT overproduction, which has slightly different consequences; maybe they show that the additional ingredients in ayahuasca are colouring the experience. Most likely it has quite a bit to do with the very different settings in which the two types of experience occur. But clearly the two share generic similarities, if nothing more. In the language of attractors, it could be proposed that ayahuasca tends mostly to directly stir up the dynamics of the Level 1 landscape, while NDEs often appear to be mainly under the control of attractors in the Level 2 landscape, producing more ordered Level 1 activity.

Despite all the similarities, there is at least one surprising difference between the two types of experience. Ayahuasca users almost invariably see hallucinatory animals at some stage, while NDEers rarely do. 'That's only natural' you might reasonably say, 'there aren't usually many animals, not even cockroaches these days, in hospital emergency rooms. But there are lots in the jungle. People are surely more likely to hallucinate what's really around or what they expect to see.' Fair enough! The only problem is that, over recent years, ayahuasca use has spread from the jungle to apartments in European cities among other places. And naïve users in these entirely urban settings, most of whom have never set foot in South America, are still liable to see animals.

The commonest animals to be seen are snakes. That is natural enough, perhaps. Most primates, including most of us, have an innate fear of snakes and can thus be regarded as having a 'snake attractor' built in to the potential dynamics of their brains. So, when the Level 1 landscape is stirred up by the brew, it's not necessarily surprising that a 'snake attractor' should be so often activated. The next most common animals to appear are black pumas, followed by jaguars. And that *is* surprising. Both are quite rare even in the Amazon, and rarer still in Europe. One could argue that South American natives have cultural preconceptions about the importance of jaguars in particular that may predispose to visions of them, but this doesn't explain why *Europeans* should see them in preference to similar animals more likely to have a firm place in their visual memories (that is, leopards, cheetahs, etc). Why should this be so?

There's no easy answer to the question. My own feeling is that the black pumas, and especially the jaguars, seen by European drinkers are so counter-intuitive that they must be hinting at something profound about the nature of attractors. We do have an intimate, hard-to-understand relationship with cats in general, so maybe it would not be so surprising if people saw ordinary cats. But they don't, they see big cats. Maybe any animal seen under the influence of ayahuasca has to be big (the snakes are often giant ones), so their size alone might not be so very odd. Perhaps, it could be further argued, the cats have to be black as well as big, through some physiological quirk or whatever — and black pumas are the only big cats to fill that particular bill. Jaguars, on the other hand, are spotty black and white, so why don't Europeans see leopards in preference to jaguars?[2] After all, most Europeans will often have seen leopards at the zoo or featured in nature programs. Jaguars get a lot less publicity.

It's hard to be sure what this oddity may mean. I think it best taken as an indication that attractors are, or can be, more law-like and transcendental that we readily allow. Ayahuasca tends to activate 'puma or jaguar attractors' in drinkers, and apparently will do so whatever the detail of their personal histories. But this attractor, unlike the snake one, has no obvious grounding in our evolution or heredity. Given our African ancestry, we might understandably harbour a predisposition to acquire 'lion' or 'leopard' attractors.[3] But no evolutionary mechanism could have given Europeans specific 'black puma' or 'jaguar' ones. As Jung repeatedly told us, that sort of archetype/attractor appears to have its home in some far more abstract realm. All the same, it can mould the hallucinations of real people. We need to move on and look at a range of other types of attractor and their landscapes before hazarding any guesses about the nature of this 'abstract realm,' or how it could influence the attractor dynamics of a person's brain.

5. Of Aliens and Mesmerism

Thus far in the book we've seen how attractor landscapes are implicit in the genetic code (briefly, in Chapter 1). Then we took a look at their embodiment in neural nets, where long term memories play a similar role in relation to functional landscapes as genes do in relation to the landscape which guides the development and physical abilities of organisms (Chapters 1 and 3). We also saw how 'Level 1' attractors in neural nets, which are emergent properties of the interactions of neurons, themselves create an emergent dynamic, manifesting in 'Level 2' attractors. These, too, are preserved as potentialities by long-term memory. In Chapter 4, we found hints that attractors in the brain may be more law-like than is generally supposed, in that they can sometimes appear to possess a supra-personal, transcendental quality in circumstances where this can't easily be attributed to underlying genetic, biological or even environmental causes. The next step is to see what other types of attractor landscape there are which may influence our minds, and begin to think about how the different sorts of landscape can influence one another.

Attractors may, in principle, exist in any dynamic system, and we need to try to get to grips with some of those to be found in society, where the relevant systems are made up of interactions between individual people, rather than the individual neurons on which 'Level 1' dynamics are directly based — and 'Level 2' are based at only one remove. Attractors in social systems ultimately derive from neuron dynamics, since brains are what create societies. But anyone might be forgiven for thinking that the connections, let's say between a neuron in someone's frontal lobe and the latest hot fashion from Milan, are often separated by so many intermediate stages that it would make sense to ignore them, just as it usually makes sense to ignore the fact that physiology is ultimately down to quantum dynamics. However life is not as simple as that, because attractors not only emerge from, but also appear to influence the behaviour of the dynamics of each lower level.

There are fewer reciprocal influences in ordinary physical systems than in attractor dynamics. As physicists like to point out, chemistry is 'simply' quantum physics in action, but lighting the gas doesn't obviously affect the sub-atomic particles of which it is composed. Similarly, the fact that some chemical compound is playing a part in my metabolism, does not affect its own structure or potentialities. The nineteenth century 'Vitalists' thought there was something special about the chemistry of life, but we now know they were wrong. Or at least they were wrong when they supposed that the *rules* of chemistry were any different in living organisms. Apparent differences are due only to the huge complexity of life compared to non-living systems.

'Level 2' dynamics, on the other hand, as well as emerging from 'Level 1,' also influences how Level 1 can and will behave. And these top-down influences are rather unlike anything seen in ordinary physical systems. To be sure, hydraulics — going back to the old example — may appear to affect the behaviour of water molecules, but it does not affect their basic structure or properties. Level 2, on the other hand, seems to do just that in relation to Level 1. Level 2 attractors that emerge in the context of NDEs or the ayahuasca experience appear to encompass a law-like quality capable of affecting the very structure of the lower level. Or maybe Level 1 changes, due to DMT or whatever, are primary and alter the Level 2 dynamic which then feeds back to Level 1. In either case, there is no such intimate feedback between liquidity and water molecules; only feed forward from molecules to liquidity, which is limited to affecting statistical properties of the large-scale movement of molecules.

As far as attractors are concerned, there's no reason to think that reciprocal relationships should be any less close as one goes further up the dynamic hierarchy. Even the big step, from brains based on interacting neurons to social systems based on interacting people, might not be any exception since top of the hierarchy brain dynamics is already, as we all know, mainly about social things — education, work, sports and all that. Because of reciprocal interconnectedness, it is likely that any hierarchy of dynamics won't be much like the relatively linear one stretching from fundamental physics to physiology. It's likely to have important loops and bridges which will give it some at least of the characteristics of a small world network — the famed 'six degrees of separation': a catch phrase deriving from the fact that in many networks (biological, neural and social), you can get from any one randomly selected point to another via a surprisingly small number of intermediate steps. It's said, for example, that you can probably get from Joe Bloggs in Lancashire to the president of Outer Mongolia via a chain of only six people who are acquainted with one another.

This picture implies that the entire system, from neuron to Milan fashion show and more, might be sufficiently integrated to embody its own top level attractors. Now there's a thought! — Jungian archetypes envisaged, not as features of human biology nor as shadows of realities existing in some ideal Platonic realm, but rather as emergent properties of the biology and the social interactions and the knowledge and aspirations of all of humanity. But it's best not to get too carried away. A concept like that is definitely 'castle-in-the-cloudish'! We need to look at lower levels in the hierarchy in order to have any hope of getting a clear impression what may be going on.

I want first of all to describe just a couple of attractors that appear to have emerged from social dynamics. They will occupy the rest of this chapter. My first example is the experience of Alien Abduction, chosen because it links into one of the main themes of the previous chapter; thus showing links between neuron dynamics and social dynamics in a fairly 'in your face' sort of way. The second is Mesmerism, which I picked mainly because it is complex, dramatic and generally a fun topic; more so, perhaps, than many comparable social phenomena. So ... on to the aliens:

Alien abduction

When Rick Strassman injected his volunteer subjects with DMT, most of them reported visual hallucinations of various types, plus OBEs, emotional experiences or sometimes even mystical experiences. A few, however, encountered aliens. In one case these were 'elves,' holding up complicated placards and able to control the subject as they wished; in another it was a sinister 'insectoid' harbouring hostile intentions; in a third, the aliens were 'flexible, fluid, geometrical cacti,' who inserted probes into the subject. Similarly, ayahuasca drinkers sometimes see extraterrestrials and spaceships, as well as the more usual jungles, animals or mythical scenes.

People with schizophrenia don't usually have vivid visual hallucinations; they tend to be more bothered by 'voices.' However, they quite commonly feel that they are being controlled by machinery implanted in them by extraterrestrials or sometimes by alien telepathy. In early textbooks, they are usually described as attributing feelings like these to supernatural agencies rather than aliens. Times change. It's true that even nowadays ayahuasca drinkers are more inclined on average to see divine figures and angels than ETs, but that's understandable as many of them take the brew in overtly religious contexts.

It certainly looks, therefore, as though a propensity to 'encounter' aliens of one sort or another is built into the dynamics of our brains, and can be brought to light by drugs or mental illnesses. Spaceships, too, have a much longer pedigree than is often supposed. There was the prophet Ezekiel's famous sky chariot or ship, with its 'wheels within wheels,' appearing in Old Testament times. The Middle Ages were quite often troubled by UFOs (Unidentified Flying Objects), then sometimes attributed to witches passing on their broomsticks. There was a major UFO scare in San Francisco and neighbouring areas in early 1897, due to frequent, inexplicable 'airship' sightings. People have evidently always been susceptible to experiences of this general sort. But the details and frequency of the experiences involves social dynamics feeding back to the brain dynamics. The potentiality to have them may be inbuilt in many or all of us, but the 'story' that activates and moulds them is a social construct or, in the terms of this book, an attractor emerging from the relevant social dynamics.

A whole lot of different threads went into the making of alien abduction. Scattered cases — people stolen away by fairies, for instance, or becoming aware that whole chunks of time were mysteriously missing from their memories — have always occurred and caused temporary, local stirs when reported. But the scene for the epidemic of abduction that happened in the later 1970s and the 1980s was set by the immediately post-World War Two flying saucer mania. Discoid UFOs had been seen earlier and, in a 1920s film, Flash Gordon had travelled to Mars in one of these. And the theme of invasion by ETs had long been popular, thanks mainly to H.G. Wells' *War of the Worlds.* Then, in 1947, a sensible private pilot saw nine disc-shaped craft one day, which appeared to be travelling at more than 1000 mph. He made some comment to a newsman about them looking like saucers skipping across a pond and the 'flying saucer' was born. Many more were seen later that year and in following years by sober, reputable witnesses. Special commissions were set up to investigate them, but few firm conclusions were reached. Pulp science fiction stories proliferated, many incorporating an abduction theme along with a flying saucer one. Then psychiatry got involved.

One evening in 1964 a moderately successful New York artist named Budd Hopkins was driving with his wife and a friend from their beach house to a cocktail party. It was an upmarket, Cape Cod affair. There had been much discussion of UFOs in their circle recently, which had been seen by several acquaintances. On the way to the party, they themselves saw in the sky what could only have been a UFO. Hopkins told the press of the sighting and got a good deal of publicity. Lots of people contacted him about their own UFO experiences. Some involved encounters closer than mere sightings of objects in the sky. The whole thing escalated. Hopkins became ever more intrigued as the years passed, as did many other people; he was the tip of an iceberg, but an important tip as he was a talented publicist and a respected figure in New York society.

The theme of interference or abduction by aliens cropped up ever more frequently in the stories that people told. Later on, in the 1980s, there were estimates that 2% or more (over four million) Americans had probably been abducted at some time in their lives. Those aliens were certainly busy!

Before abduction became quite as widespread as this, or at least before it was known to be so common, Hopkins had developed a theory that anyone who had unaccountable emotional problems and a history of 'missing time' might have been an abductee. Such a traumatic event could well have caused people to suppress all memory of it, he surmised. But nevertheless the event left its aftermath in emotional disturbance. A psychiatrist friend agreed to hypnotize carefully selected cases of this sort, to see whether repressed memories of abduction could in fact be recovered. This proved to be the case, at least in Hopkins' opinion. The friend was more sceptical about the reality of the memories. So Hopkins learned hypnotic technique himself. He seems to have been motivated by a wish to help these cases with their trauma, as well as a desire to discover more about what aliens got up to. And other, sometimes less cautious, psychiatrists got in on the act ...

A fairly consistent story line soon emerged from these 'recovered memories.' Most aliens were three or four feet high with greyish skin, large tear-drop heads, no ears or hair, big dark eyes and lipless mouths. Some witnesses reported other varieties, but these were a minority at first. They would zap their victim somehow, who was often a woman and frequently in bed, either alone or while her husband slept. The next thing she would recall was being in some sort of medical facility in the space ship. She would be examined and probed, most often gynaecologically. Samples were often taken from her ovaries or elsewhere. Most times, the process was painful. Gadgets of unknown purpose might be implanted in her nose or head. Then she might have a conversation, maybe a telepathic one, with an alien. Given the circumstances, these conversations were almost always remarkably banal. Next thing, she would find herself back in her own bed, feeling sore and upset and maybe with a nosebleed or a few strange marks on her skin to show. When men were involved, the procedure was much the same — except the samples were often of forcibly extracted sperm.

There were meetings and conferences and theories and schools of thought. Hopkins and his school stuck to a fairly hard line; aliens were real and probably up to no good. Others took a more New Age view, arguing that the aliens were here to promote greenness and spiritual growth. The most prominent psychiatrist of all to get involved, a Harvard professor no less (John Mack), wondered whether they might be 'real' only in some nebulous sense; entities belonging to a spiritual world or another dimension — or maybe emanations from the unconscious mind. Meanwhile the stories told by abductees became ever more bizarre; the variety of aliens and spaceships proliferated; they took to often floating people around on beams of blue light and 'independent' witnesses claimed to have seen such marvels; probable hoaxes were perpetrated, though some of the faithful always took them at face value. Star abductees became involved in what could look to the uncharitable like a competition to see who could be abducted most often. One or two got away most nights for quite long periods at a stretch. Some were even snatched from the venues of the very conferences they were attending.

Interest peaked in the late 1980s; then the sheer implausibility of it all struck home. UFOs are still seen and people report abduction, but they generally now rate only a paragraph in a local newspaper. A relatively unattractive Scottish town, for instance, was attracting a good deal of UFO attention in the early 2000s, thus demonstrating the impenetrability of the alien's motives. There are much nicer places nearby that they could have visited.

It's a wonderful story, for there are so many lessons to drawn from it. One is that it's only too easy to draw the wrong lesson. Budd Hopkins, for instance, argued that the use of hypnosis to recover repressed memories of childhood sexual abuse, which had become all the rage a bit later than alien abduction, validated his use of the technique to bring abduction to light. The correct argument would have been the exact opposite; namely that his use of hypnosis to 'recover' memories showed that the technique is capable of creating them. It's easy to be wise with hindsight. We know about false memory syndrome now, but few people in the 1980s realized that it exists.

What seems to have happened is that a group of concepts which had been floating around for a long time, some of them with a probable basis in brain 'hard-wiring,' suddenly coalesced and appeared to take on a life of their own. Budd Hopkins, artist and socialite, proved particularly susceptible. But he would surely have been dismissed as a crank and ignored, had he delivered a message not in accord with the spirit of the times. Indeed it's likely that the message itself was a creation of the spirit of the times, rather than being all his own work. That has to be an indication that a new attractor had formed in the 'landscape' of (mainly) American social dynamics.

Why 'memories' of little grey aliens? Well, that's an indication of how an attractor in the social dynamics can feed back to influence levels 1 and 2 brain attractors. The propensity to see aliens at all seems inbuilt, probably at Level 1, while the 'story' that is told about them is a Level 2 phenomenon. The social dynamics appears to have moulded both the basic perception — to produce 'greys' instead of the more traditional elves or hobgoblins — and the actions experienced together with the settings in these occurred.[1] The most obvious mechanism through which the top of the attractor hierarchy affected the bottom, in this case, was the use of regressive hypnosis. No doubt others, less direct, were also involved; especially feedbacks involving all the books, articles and conferences on the topic, which helped to disseminate and strengthen the whole thing. But there was feedback up the hierarchy, too. Levels 1 and 2 perceptions had to be fed into society before a consensus could be reached over what form they should take. And it proved an unstable consensus. The overall attractor landscape had shifted by 1990, and 'alien abduction' was getting ever shallower and less attractive.

So what can the story of Mesmerism tell us?

The story of Mesmerism

Franz Anton Mesmer (1734–1815) was a bright young man from a working class background. Born in Swabia, now part of southern Germany, he went to local, Jesuit run universities, then to the prestigious medical school in Vienna. His doctoral thesis was on what he termed 'animal gravity.' It was an universal force, a generalization of Newtonian gravity, which could affect people's well-being and the course of disease. His professors were happy with the thesis and he became in due course a successful and prosperous practitioner. Some of the prosperity derived from his having married a wealthy widow. However, he did not lose sight of his earlier ideas. As they developed, 'magnetism' took over from 'gravity,' but he realized that it was not everyday magnetism for it affected all sorts of materials other than metals. Eventually he discovered, with the cooperation of a particularly difficult patient named Miss Oesterline, that he could beneficially transmit this magnetism to her by making downward 'passes' with his hands over her body. Vienna evidently has a long tradition of throwing up 'difficult' patients with a talent for confirming the validity of their doctor's pet theories. More than a century later Sigmund Freud, for example, came across several similar young ladies.

When Mesmer tried to get colleagues in Vienna to investigate and publicize his discoveries about animal magnetism, they proved less than enthusiastic. Nevertheless he wrote a pamphlet about his ideas and lectured in neighbouring countries. He continued to produce remarkable cures from time to time with his 'passes.' Then things began to go wrong. There was a to-do of some sort over his (successful) treatment of an eighteen year-old pianist, Miss Paradis; he was on increasingly bad terms with his wife; colleagues in Vienna remained unappreciative. In 1778 he transferred to Paris, the very centre of the civilized world.

For some reason he turned out to be exactly what the *beau monde* of Paris, especially its female half, had been waiting for. So many flocked to him that he was forced to develop group treatment methods and rope in assistants to help. Patients would come and sit on the edge of a large tub, filled with iron filings and 'magnetized' water. Moveable iron rods could be grasped or touched to afflicted parts. Meanwhile assistants, or maybe the master himself, would appear in their robes and make magnetic 'passes' over the assembled sufferers. It all seems to have been a splendid piece of theatre, and pretty harmless compared to many eighteenth century 'cures' — bleedings, cuppings, vomitings and so forth.

Unfortunately for Mesmer, though, it was not totally harmless. He believed that his methods promoted the normal flow of 'magnetic spirits,' which had become obstructed in his patients and thus caused their symptoms. However, the process of unblocking the flow could result in a 'crisis,' rather as breaking a logjam can cause a flood. Many patients took to this idea with enthusiasm, and soon were having crises at every magnetic session. But, to disapproving husbands, their cries sounded distinctly orgasmic. And, of course, cure often arrived as (male) assistants were making their passes. It was all too shocking. Mesmer had to be discredited, but without mention of the sexual angle. The Parisian medical establishment was happy to help with the process. No-one likes a successful outsider, especially when he earns huge fees that might otherwise have entered one's own pocket.

Two royal commissions were set up to investigate Mesmer and his claims. The first had both medical and scientific members; Benjamin Franklin, ambassador in Paris at the time, was co-opted in as one of the latter. Another was Antoine Lavoisier, the great chemist. Membership of the second commission was drawn from the Royal Society of Medicine With one dissenting voice, it produced what amounted to a straightforward denunciation of unorthodox practice.

The proceedings of the first group were much more interesting. The commissioners themselves carried out a series of quite sophisticated experiments on the effectiveness of 'magnetic' treatment. They recognized that 'imagination' played a large part in bringing about any apparent cures and proved fairly conclusively with 'blind' (but not double-blind) trials that no magnetism or other physical factor was responsible. They ended up issuing vague warnings about potential dangers of the 'crises' and even hinted, albeit only in a separate, secret report, that these dangers consisted mainly of risks to sexual morality.

The non-secret reports were rapidly printed and circulated, and a war of words ensued. Mesmer had attracted disciples and imitators as well as grateful patients. There was no shortage of people willing to support the magnetic cause. Indeed most of the action now transferred to these followers. Mesmer himself took to travelling a good deal. Then the French Revolution came, causing many practical problems. He eventually moved back to Southern Germany, where he remained an active and apparently effective magnetizer into old age. He was even, towards the end of his life, accepted by the German medical establishment.

Meanwhile other Mesmerizers, as they were beginning sometimes to be called, had taken centre stage. One of the most prominent of these was Amand-Marie-Jacques de Chastenet, Marquis de Puységur, aristo, colonel in the artillery and, it appears, natural-born healer. He also seems to have been a thoroughly decent man, who would often treat the poor for free. Unlike so many aristocrats, he survived the revolution, dying aged 74 in 1825. He developed a variant treatment in which there was less emphasis on 'crises' but more on 'somnambulism,' which seems to have resembled what we would now call an hypnotic state.

A school of magnetizers, founded in 1784 by a surgeon named Pierre Orelut, developed and flourished in Lyons, which liked to regard itself as France's second city. They too developed somewhat distinctive methods, including magnetization at a distance. These people and many others had their own followers. De Puységur's disciple, Joseph Deleuze, became particularly well known and was mainly responsible for re-establishing mesmerism in Paris during the 1820s. It had fallen into disfavour there for a time, perhaps because many of its enthusiasts either did not survive the revolution or had had more urgent things on their minds. At around the time of the Paris revival, Germany too became an important centre and added its (then) typical admixture of romanticism and mysticism to the theory of mesmerism: a topic already overly obscure, so rationalists felt, without a Germanic contribution.

Twenty years on, the 1840s saw Mesmerism's heyday in Victorian England. Lords and ladies, natural philosophers and clergymen, burghers and doctors all discussed it, while many themselves made experiments. Magnetizing lecturers and demonstrators traveled the country, showing the wonders of their art. Writers and cartoonists frequently made use of it, to the extent that mesmeric 'passes' and other mannerisms became clichés. And it had very practical benefits, too. People were sometimes cured, or at least symptomatically improved, by mesmeric treatment. De Puységur's 'somnambulistic' states provided surprisingly reliable anaesthesia at a time when chloroform, and even ether, were lacking

Historian Alison Winter wrote a delightful account of what went on. She commented:

> Mesmeric seances were certainly frequent, even everyday, events. But the Victorians who attended them recorded a fascinating, disturbing, and sometimes even life-changing experience ... The mesmerist demonstrated the essence of influence; the subject displayed amazing feats of perception and cognition.

The whole topic had moved on, in other words, from being nothing more than an unorthodox medical treatment to an exploration of the nature of corporality, mentality, spirituality and all the rest. Some mesmerizers could transform their subjects into 'living marionettes,' in much the same way as twentieth century stage hypnotists induced people to behave as 'robots' or whatever. Others established what we would now call 'telepathic' links with them; or 'released' subjects' spirits to go on journeys to distant lands or into the future, as if in a strangely distorted revival of shamanism.

To a society just beginning to get to grips with electro-magnetism, the notion of 'animal magnetism' made some sort of sense, even though the educated were aware it had nothing to do with the ordinary variety. A fluid essence of consciousness that could be transferred and shared between magnetizer and subject appeared no more improbable to many than the invisible lines of force which Mr Faraday had made manifest with his iron filings. And mesmerism produced marvels that eluded genuine electro-magnetism. Subjects could describe what their magnetizer was tasting, viewing or thinking about, even when apparently unable to see or hear him (it was usually, but not always, 'him'); they felt irresistibly impelled to copy his movements; they were oblivious of hot coals applied to their skins, of needles rammed under their finger nails.

Why then did mesmerism lose its appeal? A generation on and it had all but disappeared from mainstream culture, eking out a tenuous existence in backwaters. Its main heirs, hypnotism and spiritualism, were pale shadows of particular aspects of the magnetism that had flourished in the 1840s. Alison Winter argued in her book that a major reason for decline lay in its entanglement with Victorian debates on authority, whether social, sexual or racial, where it often displayed a covertly subversive influence. Just as in Mesmer's case, doctors in particular played a significant part in its downfall, she suggested. The 'magnetic mania' coincided with a period of medical reform, when the prevailing ideal was established that patients should be patient. They should be submissive bodies arranged silently in wards, waiting on the wisdom and other ministrations of their surgeons and physicians. Although mesmerized subjects might *seem* ideally submissive, experience soon taught that this appearance could be all too illusory. Using passive, indirect manipulations, it was often the subject, not the magnetizer, who was in real control of a mesmeric session. Doctors took umbrage at this, Winter proposed, and set out to eliminate a method so subtly subversive of their authority.

In support of her argument, she pointed out that, when ether arrived on the scene, it and mesmerism were fairly evenly matched in anaesthetic effectiveness (on average, some individuals might do better with one, some with the other) and took about equal time and skill to administer. Ether, however, had the apparent disadvantage of killing some of its recipients. There was no obvious, rational motive behind the medical preference for it. Ether proved more acceptable because it better fitted the type of social relationship with their patients that surgeons desired. And if patients died in the cause ... well, the statistics of the time were primitive so they were easily forgotten.

Despite all this, it is doubtful whether doctors should be thought wholly to blame for the principle responsibility for mesmerism's downfall. Many British ones had been enthusiasts, most prominent among them John Elliotson, a well respected physician and professor at the

University of London. The most vigorous opponent of his efforts to spread the mesmeric word was Thomas Wakley, editor of *The Lancet* and scourge, not member, of the medical establishment. Apparently as a consequence of Wakley's anti-mesmeric vituperations, Elliotson, who *was* a senior establishment figure, was forbidden by the governors to use or demonstrate mesmerism in his own hospital. This happened in December 1838, *before* 'magnetic mania' had really taken off generally in Victorian England. Wakley was a slightly earlier version, in a more limited field, of Thomas Huxley, that great evangelist for evolution and promoter of professional, 'rational' science. The 'spirit of the times' favoured people like that, and doctors were as much at its mercy as anyone else.

It's worth emphasizing that rejection of mesmeric treatment had nothing to do with its actual effectiveness relative to other contemporary methods. One place where it did get established for a short time was Bengal, thanks to the efforts of a Scottish surgeon employed in the colonial service, named James Esdaile. He discovered that mesmeric anaesthesia was a hugely useful adjunct to surgery, and managed to get the Lieutenant-Governor responsible for the province on to his side. Following the inevitable committee of enquiry, a mesmeric hospital was established in Calcutta in 1846 for a trial period of a year. Esdaile was required to test mesmerism's medical as well as surgical benefits, and to learn whether class and race affected outcomes. Although systematic comparisons with other treatment regimes don't appear to have been made, everyone — outsiders as well as staff — seems to have thought the results impressive. Anaesthesia was the star turn, though a variety of astonishing cures were also reported. If there was unease, it was down to the fact that Esdaile had trained Indian as well as European assistants in his art. And how could it be proper for an European to submit to being mesmerized by an Indian? In fact this worry seems to have been more theoretical than practical, as the vast majority of patients were Indian.

At the end of the probationary year, the hospital was declared a success. But a change of Governor and a swing to right-wing 'Indians should help themselves' views led to loss of financial support and, in 1848, it had to close. However it re-opened that same year due to popular demand and through local fund raising, but on a smaller scale. Relatively little mesmeric work was carried out, allegedly due to a shortage of mesmerists, and the institution fell slowly into decay. Esdaile himself returned to Scotland in 1851, due to ill health. A successor, Alan Webb, was appointed and there was a brief surge of renewed activity, but it soon fizzled out.

Ether took over; mesmerism faded from memory or was subsumed into 'hypnotism.' This was and is a term used for what amounted to mesmerism's 'somnambulistic' aspects, which had been coined by another Scotsman (James Braid) in 1843, though it only gradually came into general use. Perhaps its eventual popularity was due mainly to the fact that it was different; it was 'scientific' and didn't imply any of Mesmer's eighteenth century flummery or subsequent Germanic mysticism. Actually, it wasn't scientific (hypnotic states have nothing to do with sleep and the term has caused much confusion in consequence), while it's likely that a lot of the effectiveness departed along with the flummery. By all accounts, mesmerism worked better than twentieth century medical hypnosis. When flummery returned, Svengali-like in the 1890s, it had more to do with stage acts than with medical treatment. By the time anaesthetic usefulness had been rediscovered by clinicians, chemical agents were so much more effective, reliable, easy and safe to administer that there was no longer any real contest, except in a few specialist areas such as childbirth and dentistry.

What can be made of this story? Overall, it seems a fairly average tale. A craze arises, proliferates and then fades; it happens every day. Even the time scale of around three generations is middle-of-the-road. A few of these phenomena, religions for example, last a thousand years or more. Others, like yo-yos or hoola hoops, come and go in half a decade. On the other hand, this particular story is a bit special in that mesmerism's background dynamic landscapes are probably easier to imagine than most.

Before Mesmer got his big idea, educated eighteenth century people had acquired a number of newish notions — Newton's universal gravity; the 'vital spirits' of Descartes and others (that is, the working fluid powering a body and brain pictured as an hydraulic machine); plus the magnetic fluid or flux manifest in compasses and the like. These comprised attractors, or valleys in their mental landscapes. They can be pictured as having quite small 'attractor basins' (see Chapter 2). There was another attractor with a much larger basin, too, a huge one that exists in all our mental landscapes: namely the predisposition to seek out cures for our ills. In Mesmer, the working-class lad made good, this landscape underwent an upheaval. Its separate features coalesced and deepened in consequence. Because similar features already existed in contemporaries, the change that he underwent proved attractive in the everyday sense to many of them. On the other hand people who had already invested a lot of time and energy into moulding their landscapes into incompatible configurations, especially professionals, were naturally resistant to change.

As time passed, a social dynamic came into play, and fed back to affect the form of Mesmer's landscape. Everyone tended to include their own idiosyncratic contours and potholes in the new 'scape,' and some of these were distributed to all through social channels. The overall effect was continually to widen the 'basin' of mesmerism, so that it eventually came to include all sorts of things that had little or nothing to do with cures. And therein, no doubt, lay its downfall ('infill' would be more apt!). For other valleys, wider and deeper, were developing in the social landscape, while mesmerism had extended so far that it could no longer rely on the only permanent feature available to it (that is, the wish for cure). Its basin simply disappeared as a recognizably separate feature, even though many contours of newer landscapes were affected by what they had absorbed.

Someone might object that interpreting the Mesmer story in terms of landscapes is a pretty fanciful and contrived sort of thing to do. But could it point in a useful direction? That is what matters. I need to show why it might do just that and let you make your own judgments. First of all, it's back to archetypes.

6. In the Footsteps of Jung

The reason for going back to archetypes, specifically Jungian ones, is that Jung himself, widely admired in his lifetime for erudition and perceptiveness, spent a lot of energy thinking about them and tracing the roles that they play in our mental lives. Though less popular now than it was, we should not dismiss his thinking as too 'old hat' to be of value. He explored avenues that have attracted relatively little attention in the last thirty years, so may well still have something to teach us. It was pointed out earlier (Chapter 1) that Jung's idea of archetypes corresponded to that of attractors shared by the whole, or at least the majority, of humankind. He's also renowned for his concept of the 'collective unconscious' — the realm in which his archetypes had their being, he said. This was, in his view, the common property of mankind, its existence demonstrated by the 'archetypal representations' that emerge from it into the ordinary consciousness of people everywhere.

He was fascinated by the anthropology of his day, and particularly by curious commonalities that occur between ideas in different cultures. It must seem odd to all of us, for instance, that a tale identical in all essentials to that of the Pied Piper of Hamlin is said to have existed in pre-Columbian Mexico. Similarly, the greatest musician in the world was torn to pieces or flayed alive, Orpheus-like, in the stories of many cultures. Sometimes he played the pipes rather than a lyre, but the basic theme has surfaced in many places that lack obvious cultural links. Elvis Presley or Mick Jagger, of course, risked a similar fate at the hands of their teenage fans, though it was only their shirts not their skins, that were torn from their backs. They were two of the most recent embodiments of the Orpheus archetype. Jung himself thought he had proved that there was something spooky about coincidences between stories like these occurring in widely separate times and places, though his 'proofs' would not stand up to modern scrutiny.

He liked to say that every quality or archetype is accompanied by its opposite, often in the guise of a 'shadow self.' Men, for instance, harbour a female side, the 'anima,' which can cause all sorts of mayhem if ignored. Women, of course, have their 'animus'; Jungian psychologists had a wonderful time discussing its manifestations in the days of strident feminism. Jung certainly showed a similar 'complementarity' in his own intellectual life. We've already noted how he swung between viewing archetypes as built in to our biology and being expressions of abstractions existing in some sort of ideal, quasi-Platonic realm. He also had a poor opinion of groups and group mentality, which can seem odd when put in the context of his promotion of the wonders of the 'collective unconscious.'

One can empathize, if not fully sympathize, with his distrust of groups. Very much a man of his period in some ways, he thought that 'primitive' mentality lacked individuality: 'The further back we go in history,' he wrote, 'the more we see personality disappearing beneath the wrappings of collectivity. And if we go right back to primitive psychology, we find no trace of the concept of an individual.' This is nonsense, of course, though reminiscent of more recent proposals, still taken seriously in some circles if less popular now than they were twenty years ago. These include the ideas that Homeric Greeks were not self-conscious individuals or that creatures lacking our sort of spoken language must also lack conscious awareness. Notions like the 'ancient peoples weren't conscious' one are based on extrapolations from extremely sparse records. The fact that we have no Mycenean equivalent of the very personal, highly conscious, account provided by the Dark Age historian Bede, for example, is not evidence that self-aware people did not exist. It shows only that records are fragile and that the few people could write were mostly occupied on socially approved business. Social approval was mainly reserved for supra-personal concerns to do with clans, battles, royal administration and so forth.

Whatever Jung's confusion about the status of 'primitive' mentality, one can certainly sympathize with another reason for his hesitations. Being Swiss, much of his adult life was spent in the shadow of Nazi Germany. Long before World War 2, he was writing dark prophecies about how the German people had been taken over by 'the spirit of Wotan.' This spirit, he surmised, was due to their defeat and enforced conversion to Christianity by the Roman Empire. Actually, he was playing fast and loose with history there too, since the bulk of the ancestors of modern Germans had defeated the Romans, rather than vice versa, while their conversion to Christianity was mainly the work of missionaries from Ireland and Yorkshire. However, his image of the Nazi armies as legions terrorizing Europe, who had reverted to giving the Roman salute (the 'Hail Caesar' gesture is thought to have been very similar to the 'Heil Hitler' one), was apt enough. And he was quite right to point out that anyone who had prophesied such a development thirty years before it happened would have been laughed out of court.

'Any large company,' he concluded, 'composed of wholly admirable people has the morality and intelligence of an unwieldy, stupid and violent animal.' So much for democracy, then! But Jung certainly had a point. It is lucky that what we call 'democracy' is nothing like what its Athenian originators meant by the term in classical times. It was the Athenian 'democracy' which put Socrates to death, and had as much in common with Pol Pot's Cambodia as with Westminster or Capitol Hill. Despite the political rhetoric, we don't live in democracies. We actually live in what might be termed 'representative oligarchies' — as indeed, in a rather different way from us, did the Athenians themselves throughout most of their glory days — and our lives are all the better for it. Nevertheless we do have a problem. Modern bureaucracies often seem precisely to fit Jung's characterization of a 'large company.'

Jung was also fascinated by Indian religions and mythologies. Indeed his Near Death Experience was to do with a 'Hindu temple' encountered in outer space (Chapter 4). But that did not prevent him from arguing, apropos the 'growing impoverishment of symbols' in Western societies: 'If we now try to cover our nakedness with the gorgeous trappings of the east ... we would be playing our own history false.' He seems to have sought consistency, with part of his mind at least. He distrusted denizens of the 'collective unconscious,' which is in his opinion, peopled by potentially dangerous monsters, all the more dangerous when unfamiliar. Taming them involves bringing them into the light of consciousness, he believed — and it's hard to argue against this view.

One can see, in a general way, that his idea of a 'collective unconscious' had some sort of validity. It was a stronger notion than the usual concept of 'culture' because it could overwhelm and reduce people to some common denominator; often in Jung's view the lowest one. Moreover it harboured all sorts of strange phenomena, many of which appeared quite alien and numinous when they emerged into the light of day. The 'greys' of Chapter 5 are mere shadows of the terror and force possessed by the archetypal Mothers, Gods, Tricksters and so forth that Jung studied. All the same, he had major difficulties in envisaging the sort of existence that could be attributed to his collective. He knew that it must possess at least a quasi-independent existence of its own, but could never adequately formulate its basis.

At first, he thought of the collective unconscious as something 'inborn,' inherent in our biology and genetics. Later, he either avoided committing himself to any particular view or hinted that it might be grounded in some deep reality, the *unus mundus,* which underlies our semi-illusory world. Incidentally, the *unus mundus* notion is not so wacky as it may sound. Many philosophers still hold that an 'ontological' realm exists which generates, but is not the same as, the 'epistemological' reality which is all that we can perceive. Physicists busy with string theory or their colleagues interested in loop quantum gravity, assume that this picture is true. The main difference between Jung and modern physicists is that he envisaged the *unus mundus* as encompassing mentality as well as physicality. Some physicists would agree with him, but not all.

However this may be, the 'inborn' notion won't work as a complete explanation for the existence of a collective unconscious for the same reason that critics of reductionist neo-Darwinism liked to use when they wanted to needle their opponents. They liked to point out that there's no way the fine detail of brain connections can be encoded in our genes, because the amount of information needed to specify connections precisely would require roughly the number of particles in the visible universe for its representation, even if it were to be encoded at the rate of one bit of information per particle — and there just aren't that many particles in our DNA, of course. Any claim that the whole of the 'collective unconscious' could be specified by our genes runs into the same problem. There isn't enough storage space in genes. On the other hand, some individual 'archetypes,' that of the Mother for instance, do seem to be 'hard-wired' into the brains of infants. So there's probably something in the explanation, even though it is not complete.

We've already seen that in fact genes *don't* specify bodies (or brains) precisely. They merely influence the shape and content of the attractor landscapes that are inherent in the dynamic processes of growth and development. So could Jung's 'collective unconscious' be nothing more than a term for part of the overall, genetically influenced, landscape that is relevant to the development of our brains? Well, that seems a reasonable view, though maybe not the whole story. Jung was interested mainly in very large-scale features of the landscape. He would have regarded Mesmer as embodying the archetype of the Healer, for example, which is common to all cultures, and would not have been so interested in the detail of notions about magnetic flux, vital spirits and the like. And these large-scale features probably do often have a genetic basis. They are predispositions to have certain perceptions and maybe to interpret the perceptions in particular ways. They are little different in principle from our predisposition to notice snake-shaped objects and to interpret them as scary, which certainly has a genetic basis.

Nevertheless, direct identification of Jung's collective unconscious with a purely genetically moulded landscape is not quite right. For it manifests in a *cultural* context. Better to regard it as the large-scale topography of the attractor landscape at the top of the hierarchy. This landscape emerges from the dynamics of others further down, the bottom level being the genetically influenced one. And its biggest features, its great rift valleys, may well mostly reflect the genetics. But some features may emerge from what goes on at intermediate levels. The top level, too, has its own emergent dynamic, dependent on contributions from all lower levels, which feeds back to influence what goes on lower down.

Jung was primarily a therapist. After medical training, he started work aged 25 (in 1900) at the Burgholzli psychiatric hospital in Zurich. Eugen Bleuler, originator of the term 'schizophrenia,' was his boss. Then he became heavily influenced by Sigmund Freud, though they did not meet till 1907. Jung went into private practice in 1910 and was very much Freud's 'blue eyed boy' for several years. However, they grew apart apparently mainly because of contrasting approaches to human nature. Freud's deterministic picture, where everything is due to libido and how it is channelled from infancy on, simply didn't fit Jung's broader views, particularly his spiritual ones. The famous breakdown, when he encountered Philemon the spirit guide (Chapter 1), occurred during World War 1. He was not a combatant as Switzerland was neutral, but was nevertheless under a lot of psychological stress due to disagreements with Freud and trying to keep both a wife and a mistress happy — an endeavour which did not sit well with his strictly Protestant upbringing. Remarkably, whether due mainly to him or to them, the *ménage à trois* apparently survived for a long time and, for some of that time at least, to the mutual satisfaction of all three participants. Meanwhile he continued in his private practice and indeed depended on it right into old age, when he came in the view of many to embody one of his own archetypes — that of the Wise Old Man.

A lot of his writings, therefore, are about the detail of treating particular patients. His aim was to encourage 'individuation' — in his terms, escape from both 'primitive' group mentality and the rule, or destructive intrusion, of any single archetype. The goal was to become a balanced and unique individual, oriented towards spiritual growth and empowered, not disrupted, by the ambient archetypes. The method was to help patients explore their own psychic structure, bringing archetypes under conscious scrutiny and hopefully integration into a healthy psyche, through the examination of their dreams, artworks, life histories and so forth. All very ambitious and holding great appeal to many; indeed Jungian approaches to therapy are still popular in some circles. Like all the so-called 'depth psychotherapies,' though, it seems mainly to help the 'worried well.' Hans Eysenck, a well-known psychologist of the late twentieth century, cynically commented apropos of another 'depth' therapy that the ideal candidate for Freudian psychoanalysis was 'young, attractive, intelligent and, above all, perfectly well.' It was of no use, Eysenck implied, for people who had any serious mental disorder. And there was a good deal of truth in what he said. Jung's therapy tended to attract the middle-aged and those who, a generation later, would have become 'new age' spiritual seekers, but was probably no more effective than psychoanalysis as a medical treatment.

Nevertheless many experienced it as beneficial. One patient, whom Jung described at length, was a 55 year-old American psychologist who had allegedly had a poor relationship with her Danish mother. She had never married, Jung surmised, because she harboured a particularly strong 'animus' deriving from her 'exceptional' father, and thus had no need of any external male companionship. Well, it was a theory! Anyhow, she had come to Europe in search of her maternal roots and to Jung because she had read up on his writings and found them appealing. In the course of their relationship, she produced a series of twenty-four symbolic paintings, starting with a lonely female figure emerging from some dark, egg-like rocks; progressing through a whole series of 'Mandala' images (abstract or semi-abstract symbols, enclosed within borders); culminating in a mandala comprising a central lotus flower, with a star above it and two snakes posed symmetrically and benignly beneath. The snakes had appeared in previous paintings in more threatening and disruptive roles. Psychic peace and harmony, in other words, prevailed in the end. The star, a cynic like Eysenck might say, no doubt represented an award to a top pupil.

Jung does not tell us anything definite about whether this benign outcome persisted or extended into his patient's everyday life; merely that she died of cancer twelve years later. Indeed the impression he gives is of not being much interested in the patient herself, only in the symbols and images she produced for him. And she seems to have shared his focus. When discussing a pair of wings in one of her pictures, for example, he reports her as saying: 'Naturally they are the wings of Mercury, the messenger of the gods. The silver is *quicksilver* [a powerful alchemical symbol for Jung]. Mercury, that is, Hermes, is the Nous, the mind of reason, and that is the animus ...' They probably spent a fascinating time together pursuing their shared and highly erudite obsession with symbology — and perhaps it really was helpful to her.

Much of all this comes across to us nowadays as contrived nonsense, but perhaps with a more serious, maybe even a more solid, foundation than the rather similar New Age fads that have proliferated since. It is quite difficult, now, to get an impression of the force or validity of what Jung was saying from his own writings. Too much water has flowed under the bridge since. However, the power of his vision has been beautifully portrayed in a way that can still 'get through' to us, by that very great if often neglected (particularly in the UK) Canadian novelist, Robertson Davies (1913–1995). His *Deptford Trilogy*, a series of three novels centring on a 'biography' of a schoolteacher, contains delightful, insightful portraits of an archetypal Trickster and of Jungian therapy. Having a teacher as the central character may sound dull, but isn't. The whole series is really about the 'individuation' that takes place in him and subsidiary characters, and is as enthralling as any thriller.

So, assuming his method did have some validity, what was Jung achieving with his therapy? Its essence, unsurprisingly given that Jung was a former disciple of Freud, was to attempt to tackle 'dissociation' (see Chapter 3). That's not surprising because Freud considered all neurosis attributable to this. In attractor landscape language, Jung was trying to identify covert attractors that were doing their own thing independently of the overall dynamics, and bring them into the fold so to speak. He aimed to integrate the independents into, and bring them under the control of, the dynamic that underpins the flow of consciousness. But, as we've already seen (Chapter 3 again), 'destabilizing system' and 'attentional system' abnormalities are more likely to be at the root of many psychiatric conditions than is 'dissociation' — hence perhaps the ineffectiveness of both Freudian and Jungian therapy for these conditions. Nevertheless, Jung's approach does make sense in terms of attractor dynamics. It may well have been appropriate treatment for some personality disorders and could indeed have promoted personal growth just as he claimed.

I have to admit that following in Jung's footsteps has not taken us all that far. Maybe it has helped towards refining our picture of the 'social' attractor landscape by getting us to view it in terms of a 'collective unconscious.' It has been helpful, too, to have fleshed out our concept of the more universal attractors in our minds, by looking at some of the detail involved in their equivalence to Jungian archetypes. On the other hand, although we have not made any huge advance, at least we've made a start. The next step is to take a look at a concept — that of memes — that offers the prospect of achieving greater precision in envisaging attractor landscapes. Jung helped us to see the overall topography — the shapes of the continents and what they are made of, one might say. Now we need to look at much more local features, to see how they may originate and to try to understand their dynamics.

7. The Trouble with Memes

If the large-scale topography of attractor landscapes is influenced by genetics, as the 'collective unconscious' picture suggests, it's natural to wonder whether and how memetics could be relevant, too. After all memes are to culture what genes are to brains, according to their advocates. And the concept is respectable these days. The word 'meme' has got into most dictionaries. Moreover one of their main champions (psychologist Susan Blackmore) has proposed that their own dynamic was what set up a 'memetic drive' responsible for the ever increasing size of our hominid ancestors' brains, as well as the progressive elaboration of human language. In other words memes may influence genes, she suggests. Nor is this a 'way out' suggestion, for evolutionary biologists are now beginning to accept the idea that culture may affect the direction that genetic, Darwinian evolution will take. There's an excellent book entitled *Evolution in Four Dimensions,* for example, which points out that Richard Dawkins' now traditional picture of genetic evolution (still called '*neo*-Darwinism,' but it's only new in relation to nineteenth century Darwinism) is just one of four types, albeit the most fundamental, with cultural evolution at the top of the hierarchy. Causative influences, however, run both up and down the hierarchy as is the case with our various levels of attractor dynamics.

Daniel Dennett, a philosopher who has become a prolific and influential commentator on most aspects of consciousness, is also a fan of memes. They play a central role in his concepts of what conscious awareness really is, where 'free will' comes from, and the like. I won't list all of his books here. *Consciousness Explained,* published in 1991, was the first to attract widespread attention; others, which make even more enjoyable reads if you're interested in that sort of thing, include *Darwin's Dangerous Idea: evolution and the meanings of life* (1996) and *Freedom Evolves* (2004). If you have read them, you may imagine that there is no problem explaining culture, the content of consciousness and much else in terms of the behaviour of memes; specifically the fact that they are supposed to engage in a Darwinian competition for places in our 'mental space.' But there are problems nonetheless.

In 1998, I incautiously wrote one of those 'target articles,' of the sort that some journals like to publish. The idea is that the fall guy (me in this case) will say something a bit new, speculative and provocative, and the journal editor will get others to write commentaries on it. With a bit of luck, all concerned will get educated or stimulated in some way. My article outlined some tentative ideas on archetypes, memes and possible relationships between them. And it certainly worked as intended, at least insofar as it stimulated one commentator. Indeed it worked a little too well, for it provoked her(?) to a state of incandescent rage. I've no idea who it was, as she (?) wrote under the pseudonym of 'Donnya Wheelwell.' She (?) was probably an American academic of some description. I've put in the '(?)' as I toyed with the theory — more aptly described as a dark suspicion — that 'she' was the journal's (male) editor-in-chief in disguise. I liked to picture him cycling round his neighbourhood in skirt and flowery hat, shouting rude words at innocent passers-by, then pedalling furiously on. On the other hand, one can forgive Donnya a lot for the witty title of her paper — *And the Meme Raths Outgrabe.* — with apologies to Lewis Carroll (one of his famous nonsense poems featured the wonderful line 'and the *Mome* Raths outgrabe.').

She ended the commentary with a splendid piece of invective: 'This paper has tried to show how Nunn, Atmanspacher [another of the commentary writers], Dawkins, Jung, etc., perpetuate three of the most vile, vicious, virile, virulent viral strains of Western culture: atomism, reification and phallogocentrism.' I don't know what 'phallogocentrism' means, but suspect it of being a feminist term of abuse. We'll pass over her comments on Jung, since she evidently knew little of him or his writings. For instance, she waxed indignant over my reference to mandalas as symbols enclosed in a circle, on the grounds that the contents are often complex and the enclosure square. However, over 80% of those that Jung reproduced in his books were circular and the content was quite often not particularly complicated. Donnya was thinking of a subset of Indian or Tibetan mandalas, which Jung regarded as idiosyncratic representations of the underlying archetype. Some of what she said about memes, however, was very pertinent.

Her overall objection appeared to centre on the view that it is silly, futile and altogether wrong-headed to try to reduce culture to component 'atoms,' and that is what memeticists are trying to achieve. In support of her view, she argued that no satisfactory definition of a meme exists and that it is inherently impossible to make one because the concept is vacuous. Moreover, the alleged transmissibility of memes from person to person is illusory because for instance, 'a single sentence can be interpreted in drastically different ways by different people, and even by the same person at different times ...' These arguments have a good deal of face validity. Before we can deal with them, we need to look a bit more closely at the concept of a meme.

Richard Dawkins coined the word, and used it in his widely read and influential book *The Selfish Gene* (1976). He wanted a term for ideas that replicate using people's minds for their transmission, just as genes, so he argued in the book, use our bodies for their replication and transmission. Daniel Dennett has referred to them as 'minimal units of cultural transmission,' adding that they are not basic ideas such as those of 'red' or 'round' or 'hot and cold.' Rather, they are the sort of complex notions that form themselves into distinct memorable units — 'such as the ideas of wheel, wearing clothes, vendetta, right triangle ... deconstructionism.' He was getting a bit carried away by the time he got to 'deconstructionism,' for that's an idea which few, other than philosophers and literary critics, find memorable. But you get the general drift. And nobody has come up with anything much better than this. It leaves an awful lot unsaid, and people trying to plug the gaps have gone about it in so many different ways that little other than confusion has resulted.

For instance, is a four-wheeled vehicle a meme, or is there a meme for each wheel plus one for the bodywork? Intuitively one might think that 'car' ('automobile' if you're American) is a meme, but that's purely dependent on context. To Mums doing the school run, 'car' is a single memorable unit. To a mechanic trying to repair one following a breakdown, it dissolves into hundreds of memorable units. The same mechanic, taking his family for a drive at the weekend, will probably think of 'my car' as a single unit, unless it develops some strange noise in the transmission, when it will again dissolve into lots of units. Memes, on Dennett's definition, are so context dependent as to be nothing like genes. He's lost, one might reasonably think, the whole rationale for proposing them in the first place.

Then there's the question of when is a meme not a meme? Dawkins took the view that they are embodied in our minds and in all the artefacts that we use or encounter, especially books and other records. The idea of an arch, one of the instances of a meme that he gave, might be in someone's imagination, or described in a textbook of architecture or in plain view in the church we're visiting. It's a meme in all these circumstances, said Dawkins, the criterion for memeship being that it should occupy part of our 'mental space.' This space includes our memory and all its ancillary records out there in our environment. What, then, is the status of an idea in a book that no one ever reads? It was a meme when its author was writing, and is potentially a meme again if someone were to read the book ... and in the meantime? I suppose it could be said that an unread book no longer belongs to our 'mental space,' so there's no problem. On the other hand, that makes the definition of memeship dependent on an unknowable future. After all, some archaeologist might unearth the book in a thousand years time and think to herself, 'That's an interesting meme I've just found in that old book.' A definition that depends on something that is unknowable in principle is surely very iffy indeed. Maybe Donnya was right to be indignant.

Some meme fans have tended to sweep these problems under the carpet. Dawkins himself distracted our attention with parables about how memes can behave; for instance the one about gruesome parasites that put their hosts at risk. One example that he gave concerned a parasite whose primary host is sheep and its secondary host ants. It alters the ants' behaviour so that they tend to climb grass stems towards the light, the reverse of what they would normally do, where of course they are more likely to get eaten by a browsing sheep. This spreads the parasites' genes all right, but is not so good for the ants. Memes can be to us as those parasites to the ants, he claimed.

He made our blood run cold with descriptions of that frightful bogeyman, the 'memeplex.' Groups of self-reinforcing memes get together, he said, and wreck mayhem with our psyches. Religions are obvious examples of these; separate memes involved in many religions include ideas of heaven, hell, and your likely fate if you don't believe. If the Lord don't get you then the Devil must! Each meme reciprocally encourages acceptance of the others. Many other writers have followed in Dawkins' footsteps, and have also frightened us with tales of selfish memeplexes, concerned only with their own proliferation regardless of damage done to us along the way.[1] Those excellent and very level-headed popular science writers, Ian Stewart and Jack Cohen, recently collaborated on a science fiction story *(Heaven),* which revolves around a 'malignant memeplex' that is spreading through the galaxy — though readers' attention tends to get distracted from this by the wonderful alien ecologies that they picture (Cohen is a biologist). Nevertheless, frightful stories like these are proportionate to the facts. One has only to look at the history of crusades and religious and political intolerance in general to see that Dawkins and followers have a point; *some* explanation is required for the sad and violent story of human behaviour, though it is not always clear from their accounts that memes are in fact up to the job of explaining.

Other writers have squarely faced the 'how can one adequately define a meme?' problem. Anthropologist Robert Aunger, for instance, has argued at length that the only adequate concept is to regard them as fleeting, distinctive electrical states of some sort in individual brains. 'A meme, then, is essentially the state of a node in a neuronal network, capable of generating a copy of itself in either the same or a different neuronal network, without being destroyed in the process,' he wrote. Quite a lot of meme fans would go along with him; it's an attractive idea in some respects. It deals nicely with the objection that memes are so context dependent as to be illusory, for brain states *are* context dependent. Aunger's equation of memes with particular patterns of electrical activity allows for this, but nevertheless leaves his version of a meme intact. A pattern does have an identity of its own, regardless of the place or context in which it is reproduced.

But, as a definition, this one leaves ideas in books out in the cold. And ideas in books (and other artefacts) have to be incorporated in the meme picture somehow, if it is to be of any value. Every self-respecting memeplex is embodied in literature as well as brains. Many have been around for a very long time in both of these formats — Dawkins would point to the Bible as an example. The Old Testament was collated and written down over two thousand years ago and also, until recently, existed in people's neural memories too. Any self-respecting nineteenth century divine could quote large chunks of it by heart. Books are clearly involved, along with neural states, in meme replication. Memes are nothing if they are not replicators, so it looks as though Aunger may have thrown out the baby with the bathwater — or confused his 'genotype' with his 'phenotype.' They may well be expressed in our brains in particular states of neural nets, as a gene may be expressed in the development of a particular type of neuron, but that does not mean that the neuron and the gene are the same. Equally it does not follow, from a meme's expression in some particular electrical pattern, that the pattern is the meme.

We've met a similar problem before in Jung's inability to formulate an adequate basis for the existence of his archetypes. Indeed, I think that's the same problem in a slightly different guise. So Aunger's reference to neural nets offers us a big hint about the way forwards. If we say that memes express themselves in our brains as attractors in neural network dynamics, then we're up and running! Such attractors are both particular memories and (idealizations of) dynamic states equivalent to Aunger's meme definition (that is, the 'state of a node in a neuronal network').

The meme itself can be pictured as a small scale version of a Jungian archetype, while Aunger's 'meme' is equivalent to an 'archetypal representation' in someone's brain; however, memic representations are also to be found in works of art and craft, just as are archetypal representations. Before exploring this avenue further, I want to look in the rest of this chapter at the other big problem with memes, to do with how they replicate and what gets replicated.

Richard Dawkins put their replication down to 'imitation.' By this he meant copying parents or role models, going along with what you see on the tele,' formal education and so on and so forth. Anything that serves to spread concepts, ideas, habits, patterns of behaviour between people amounts to 'imitation.' Donnya's claims notwithstanding, it's clear that ideas do spread fairly reliably by means like these. Schoolchildren do learn the concept of a 'right triangle' pretty accurately, if not always as quickly as their teachers might like. Huge numbers of people end up wearing almost identical pairs of jeans simply because the 'meme' for doing so has spread widely — comfort and practicality certainly don't come into it! The fact that people may sometimes end up with a different meme from the one transmitted to them, or no meme at all, doesn't mean that memes can *never* be transmitted reliably from person to person, which is what Donnya and others have implied.

The question of *what* is transmitted is also less problematic than my would-be nemesis argued. It has to be information. We are used to the idea of information taking different forms. In books it's patterns of ink; in hard discs it's patterns of magnetization or whatever newer format the technologists come up with; in brains it's patterns of nerve activity. The information content of the word 'phlogiston,' say, is identical whether it is written in a dictionary, in a computer memory or stored in my own memory. Its meaning may vary according to where it is sited, but that's a separate issue that we'll get to later. Going back to information, the best general definition of this is Gregory Bateson's: 'a difference that makes a difference.'

Bateson was an anthropologist and polymath who spent his last years at the Esalen Institute in California. His definition of information is not as useful as Claude Shannon's 'bit' (that is, the answer to a single yes or no question) if you are a telephone engineer or computer designer. On the other hand, Shannon's definition is very restrictive indeed. In particular it excludes any notion of meaning, and recovering meaning in the context of Shannon information can be very difficult. Sticking with Bateson's definition, it is clear that a meme is information that makes much the same difference when it is transferred from one memory to another. In other words, if one person transmits a meme to another, it will cause a pattern of mental activity to arise in the second person identical, or at least very similar, to that in its originator. If two people come across the same meme in a book, both will develop similar neural patterns in consequence. If someone transfers a meme from one book to another, the patterns of ink in the two books will be alike — assuming the writer doesn't translate the meme into a different language; translation involves meaning, which is far more complex than information alone.

Since genes can be regarded as information encoded in DNA, memes are indeed, as Dawkins said, analogous to genes in that respect at least. Other objections to the meme concept, less fundamental than Donnya's mistaken ones, have centred on the fact that their behaviour can't be strictly 'Darwinian' and that there seems to be something a bit hazy about the notion of 'imitation' as the mechanism responsible for transmission. Let's take a closer look at these objections; the Darwinian one first.

Orthodox Darwinism said that natural selection acting on consequences of entirely random variations in DNA is what's responsible for evolution. In other words, throw the dice to select a change in the information content of DNA and cook in the usual way. If the outcome is an improvement on the original recipe it's likely to spread; if the opposite, the chances are it will sooner or later die out. There's no way memes can be exactly like that, since it is unlikely that variations in them will be entirely random. Moreover, although we have no choice about what genes we inherit, we can to some extent select which memes we shall incorporate in our minds. A lot do get into us willy-nilly, through parental example, what happens to be 'in the air' and so forth. But we select others by choosing to pay attention to them. Even Dawkins' dreaded religious memeplexes partly depend on choices — to attend church, go to bible study classes or whatever.

Memetic 'evolution' thus depends on inheritance of what amount to *acquired* characteristics; the shock-horror Lamarck word! Lamarckianism was heresy twenty years ago, as far as orthodox neo-Darwinists were concerned. Memes, for many of them, were thus tainted with heresy and not to be taken seriously. Views have softened since, largely as a result of the quite recent discovery that Lamarckian inheritance is actually alive and well in biology. The so-called 'epigenetic' systems, whose main function is to ensure that cell lines breed true (that is, to ensure that skin cells give rise to skin cells and not to liver or kidney ones, say), can also cross the generations. They are responsible for some 'hereditary' diseases for example. And epigenetic inheritance is very definitely inheritance of acquired characteristics. The heresy has taken over the body of the Evolutionist's kirk, and memes are no longer out in the cold in this respect.

What about the other source of doubts; the vagueness of 'imitation' as a means of transmission? DNA can be put under the microscope and viewed in eggs and sperm; the fate of its component parts followed as they combine to produce a new embryo. Information in memes cannot be so readily identified or traced. The whole meme picture can appear distinctly cloud-castleish, when viewed from this angle. However some new findings suggest that the information flow involved in transmission of some memes may in fact be traceable. If you can follow the information step by step, you've got a much more DNA-like, down-to-earth picture.

There's been quite a bit of excitement in neuroscience circles over the discovery of 'mirror neurons.' An important type of research involves recording the electrical activity in individual cells in the brains of monkeys, and trying to relate this to what they are doing, perceiving or even thinking. The 'thinking' is inference of course, but a pretty good case can be made for its validity in some of the more sophisticated experiments. Anyhow, in 1996, a group of researchers found neurons that 'lit up' both when a monkey was performing some action and when it was simply watching another monkey, or in some circumstances the experimenter, carrying out the same action. This was a total surprise because it had previously been assumed that acting yourself and perceiving someone else acting are entirely separate functions carried out in separate brain areas. But these mirror neurons were doing both. It's since been shown that we, as well as monkeys, have them. Moreover they're not just confined to visual perception. Hearing paper being torn, for instance, has been shown to activate particular neurons that are also involved in the act of tearing.

People have speculated that these and similar mirror neurons underlie our capacity for empathy. But they're also interesting from the meme point of view. Some memes involve patterns of action — wearing clothes for example, which was one of Dennett's instances. So some of the very same neurons that are active when you dress are also active when you watch someone else get dressed. Information in these two very different circumstances is thus identically embodied (in the mirror neurons) and is likely to *be* identical. Our brains thus appear to have an automatic translation system allowing a meme manifesting in the outside world to be the same, in an important sense, as one manifesting in ourselves.

This makes the whole 'imitation' process look a lot more concrete, at least for memes that involve behaviour. There's not such a big gap as one might suppose, it appears, between one brain and another. Unfortunately, it is not yet known how mirror neuron learning proceeds. These neurons could cause Dawkins' 'imitation' only if observing behaviour is capable of stimulating them into action when the behaviour itself has not yet been learned by their owner. Although there's no direct evidence (yet), this seems unlikely to happen as a one-step process. On the other hand, it does seem likely that it could occur as a result of feedback and refinement — the usual learning process in other words. Mirror neurons might be stimulated a bit by 'viewing' actions vaguely similar to ones already in their repertory. Then the correspondence between what was 'seen' and what was performed could increase step by step, along with ever greater and more specific mirror neuron activation.

We don't know for sure how mirror neurons get to be mirror neurons and we also don't know how widespread the phenomenon is. Could it underpin acquisition of memes other than ones involving simple behaviours? It's anyone's guess. But they do provide a clear demonstration that 'imitation' is a potentially observable process, at least in principle. And that's important when it comes to arguing with meme sceptics. The other important implication is to do with meaning, not just information. Our own actions carry meaning for us, but the very existence of mirror neurons implies that some of the same meaning must attach to what we see in the outside world. Thus, acquisition of a meme can involve acquisition of meaning as well as mere information. Information on its own, of course, is a very abstract sort of concept. One cannot easily get excited about it. Meaning, on the other hand, has to do with the warm, living essence of our existence. Memes have a part to play in bringing that about.

We've seen, then, that the context dependence of memes is not a problem, but simply what one would expect given that they manifest themselves in neural activity. We've also seen that they are information packages and have taken a look at how their transmission from person to person could be quite concrete and 'DNA-like,' except in so far as it involves Lamarckian, not strictly Darwinian, evolution. All Donnya's specific objections to the concept have turned out to be groundless. If she was correct in her overall assessment about them being 'phallogocentric' — well, Donnya, that's tough, but you'll have to accept it, along with the view that it may not be so silly, after all, to look at what memes can tell us about culture.

Actually, she must really have known all this despite her protestations. For characterizing the concept of memes as a 'vile ... virulent, viral [part of] Western culture' itself depends on the concept. The idea of a 'cultural virus' is equivalent to that of a meme, albeit a Dawkinesque selfish meme. Maybe that's why she got so hot under the collar. Maybe she thought there is nothing more to memes than the selfish gene analogy. If so, she was also wrong about that. I'll try to show you why in the next chapter.

8. Memes in Action

First a quick reminder about where we have got to with memes: we have found that they are packages of information, which can be recorded in any format that we are able to access; CDs, books, pictures, and so on ... or indeed our own neural memories. But they are packages with the special quality that, when activated in our minds, they manifest as meaningful, memorable concepts. Just as Jungian archetypes are the predisposition to experience or create an 'archetypal representation,' so memes provide the information (the 'difference that makes a difference') needed to experience a memic concept. This equivalence to archetypes tells us straight away where and how they affect our minds; they have to do so by influencing our attractor landscapes. And this tightens the analogy with genes, for genes affect our bodies by moulding the landscapes that guide growth and development. It probably comes as no surprise to Richard Dawkins that the analogy with genes can be pushed so far, but it is certainly surprising to most of us. Presumably the correspondence is so close because both memes and genes are simply information packages that require bodies and brains for their expression.

Is 'meme' just another word for 'Jungian archetype'? Not entirely, is the answer. For a start, most memes are quite local. They run in families, neighbourhoods, nations, professions and so forth, but are not usually the common property of all of humanity. And most archetypes are more complex than a meme; most are more like memeplexes than single memes though a few (for instance, the Mandala) are fairly simple — *pace* Donnya! Archetypes on the whole, it can be said, generate larger and more universal features of the attractor landscape than do memes. There's quite a bit of overlap, but it's probably the case that archetypes have much of their origin in our biology, while memes are mostly consequences of our social interactions.

Which leads on to the next question: where in the hierarchy of landscapes do they have their effects? Because of reciprocal influences running up and down the hierarchy, there isn't necessarily any precise answer to this question. As mentioned previously, Susan Blackmore thinks that memes may sometimes influence things as far down as the genes themselves, and there are good reasons for going along with her view. But let's follow Daniel Dennett and say that memes are not simple ideas like that of 'red' and so forth. One can then say definitely that they do not have their main home in Level 1 dynamics (see Chapter 3). This makes sense in any case since they are nothing if not cultural phenomena. It makes sense to put them at the top of the landscape hierarchy. Doing so also suggests a somewhat less circular definition of what memes are than Dennett's 'minimum units of cultural transmission'; they can be regarded as information packages that mould the smallest recognizable features of top landscapes.

Now for the scary stuff. There are rogue memes that get embedded in our minds and wreak minor or major havoc wherever they appear. A relatively innocuous one, which appears to be gradually fading out of existence, is the belief that effective medicine should taste nasty or otherwise be horrible. This probably derives from a Babylonian or earlier source. A recognized method of treatment in those times was to give patients substances so revolting that the spirits causing disease would be forced to leave home. And around 120 subsequent generations of children were made to suffer for it with bitter potions, sulphur fumigations and the like. But notice that there is a smidgeon of truth behind the belief, which is probably why it persisted for so long. Lots of natural medicines are in fact toxins produced by plants to ward off herbivores or parasites. We've evolved to recognize many of these toxins when we taste them and they are indeed nasty.

A far more disastrous meme is the idea that torture is a good way of uncovering truths that people wish to hide. In our culture at least, this seems to have descended from a notion embedded in Roman jurisprudence. Although torturing Roman citizens was not permitted, slaves involved in a judicial enquiry had to be tortured whether or not they were willing to give evidence. Romans believed that they could be relied on to lie unless 'put to the question,' to use the inquisitor's euphemism. And of course the utility of these methods is all too apparent to those who use them. People under torture are only too anxious to spill out whatever they think the inquisitor wants to hear. Hence inquisitors quickly uncover 'facts' that confirm their deepest instincts — in particular their fearful instincts. Witches can be 'proved' to hold Black Masses and fly every night on broomsticks to meet the Devil. Conspirators in almost every generation since the 'Catiline affair' in late Republican Rome have held fearful ceremonies, often involving the blood of sacrificed children. Zionist plots of enormous sophistication threatened Tsarist Russia and right-wing Germany. And, as for Islamists today ... !

In a way, torturers are in much the same position as hypochondriacs. Every piece of evidence that an hypochondriac gathers points to the firm conclusion that they have some frightful illness. But this evidence is generally the product of their own imaginations. Torturers, on the other hand, engage in a ghastly duet with their victims to fabricate the evidence that will confirm their worst fears. And, just as any hypochondriac who waits long enough will develop a genuinely fatal illness, so sooner or later the torturer will come across a real conspiracy. The meme is not only self-validating in the eyes of inquisitors but also, just occasionally, does deliver on its promise and uncover a hidden truth. So it persists, to the huge detriment of users and their societies. Best not to dwell on what it does to its victims. There is such a slippery slope. If a little bit of torture doesn't produce what's wanted, then surely a little bit more, and a little bit more, and a little bit more?

Let's leave this depressing topic and move on to another — how memeplexes form and are able sometimes to run wild. We've already seen hints about this in the way that separate concepts or memes (that is, 'UFOs exist,' 'aliens exist,' 'they are interested in our reproductive organs,' etc) came together to produce Alien Abduction. In Mesmerism's case the memes were to do with 'magnetic fluids,' 'vital spirits,' 'healing,' 'unblocking conduits' and so forth. I want to look at a couple of other examples in more detail here because they are simply too good to omit. They also figured in my *De la Mettrie's Ghost,* so apologies to any of you who may have read that. I am, however, putting a rather different slant on them here, which may be of interest even if you do already know about them. They are Tarantism and Neurasthenia. We'll take them in that order.

The case of Tarantism

Tarantism was an overwhelming, apparently involuntary urge to dance, liable to afflict southern Europeans during a period of over four hundred years starting in late medieval times. Once started, dancers would continue for hours or days, and are said to have sometimes died of exhaustion. Tarantism is especially interesting because it was apparently due to a small number of quite well-defined memes, which were the following three notions that were commonly accepted in the fourteenth to eighteenth centuries:

- In Apulia in southern Italy there lives a small, inconspicuous spider called a *tarantula* (nothing like our 'tarantula,' therefore, which is big and hairy);
- the bite of the *tarantula* is normally fatal;
- the only cure is urgent dancing.

Although the spider initially lived only in Apulia (now Puglia), its habitat expanded later on to include much of Europe south of the Alps. This occurred soon after it was first described in print, which happened in the mid fifteenth century. It was not only the Bible that distributed written memes in those times! Tarantism, of course, accompanied the alleged spider and became almost institutionalized. Special tunes (tarantellas) were composed and dedicated bands of musicians sometimes employed to cater for it.

It's fairly understandable that these memes should have spread so widely and lasted so long, once they got going. After all any minor insect bite, if you lived in a tarantula-infested area, could have been a fatal spider bite. And who would be fool enough to test its fatality by delaying the well-recognized means of cure? Even if some brave soul had done so and survived — well, they couldn't have been bitten by a true tarantula after all, it must have been some other creature. It was always a memeplex destined for success in a society where zoology and statistical analysis were primitive or lacking. But how did it ever get launched in the first place? I cannot do better than offer a story; here it is:

Once upon a time, near the great town of Bari, there was a village where people liked to tell stories. One story that all the children knew was about the dreaded *terentella* (pronunciations were a little different in those times), so small you could hardly see it, but able to strike you dead — just like that! Even the wisest in the village did not know that this story was descended from a Roman one — though the Romans told it about an entirely different creature, a lizard they called a *stellio*. The children also knew two other things. People who live where they are often bitten by insects, tend to wake up dead (actually mainly from malaria, but of course they didn't know that then). Also they knew that if you have lots of troubles and are feeling bad, quite often having a good dance will make you feel better. And there were lots of other stories too, some of them even told in church, about how dancing could prevent or cure physical and spiritual ills of all sorts.

Hard times came on the village; many died; everyone was anxious and upset. 'Who will be next? We don't know!' Maria, especially, had been mulling over these problems. She had lost two sons and a cousin only last year, and her favourite sister just a month ago. They had died quite suddenly, for no obvious reason. Maria's attractor landscapes were heaving and moiling with all the emotional pressure. Suddenly, everything re-arranged itself; various small dents, the separate ideas, toppled into a single valley and came together. 'It's the *terentella,*' she proclaimed, 'those who wish to be saved should try dancing.' And the people did dance, because they knew her to be a wise woman; and they did feel relieved in their minds; and no-one died for several weeks; and so the memeplex was consolidated.

Neurasthenia: a Victorian invention

Neurasthenia, a Victorian invention, is a bit more complicated. People who got it experienced overwhelming fatigue for no obvious reason. Other symptoms could come and go, but weariness was the principle one. Sufferers generally remained pictures of perfect health; it was just that, if they undertook any physical or mental exertion, they could not carry on for more than a few minutes because they simply felt too tired to continue. The memes behind it were:

- weakness and fatigue are symptoms of some poorly understood illnesses;
- there are illnesses that mainly afflict particularly talented and sensitive people;
- illnesses of the above type are not 'mental';
- rest is a good cure for many conditions;
- most illness is due to environmental causes (for instance, germs, pollution, 'over-civilized' lifestyle, and so on).

The first two ideas appear to have had their origins in seventeenth and eighteenth century medicine; the next two achieved prominence in the early nineteenth century, while the last was more of a mish-mash.

Eighteenth century physicians held that too much 'civilization,' in the form of fashionable city life, could be bad for you. Early nineteenth century ones were starting to get concerned about pollution and the effects of slum conditions. The mid to late nineteenth century germ theory of disease, with all its associated discoveries, put extra oomph behind the whole idea of environmental causes.

You can see why this memeplex had potential. Suppose you feel a bit run down one day for whatever reason, maybe you start to worry that you are ill. And of course you've always been a pretty sensitive sort of person — quite a bit of talent too — so you've probably got the sort of illness that goes along with that. Best thing is to get as much rest as possible; maybe go to bed for a while. You've got plenty of time on your hands while resting and you worry some more. Quite soon you find that you are able to do less and less, however much you rest. In fact, of course, unduly prolonged rest *causes* weakness and fatigability, but you don't see that because you know rest is best. And, if you didn't know it to start off with, there were soon plenty of doctors and other 'helpful' neurasthenia specialists around to tell you.

And the other two memes? — the ones about illnesses like this being non-mental and due to environmental causes? They provided the cop-out that allowed the whole group to flourish and spread: 'It's all something that has come over me, due to some bug maybe or something out there; anyway it's not my responsibility. Those doctors should pull their fingers out and discover what it is and how to cure it.'

The whole thing finally appeared to fall apart in the early twentieth century when people lost faith in the truth behind some of its component memes, especially the one about 'non-mental' origin. But the 'plex' was tougher than it seemed, for it bounced back a few decades later under a variety of different names, none of which sounded anything like 'neurasthenia.' I won't re-tell the story of its resurgence here, full of drama though it is, as it would take us too far from the theme of this chapter.

We know the identity of the 'Maria' of neurasthenia. It was a 'him' in this case; an American neurologist named George Beard. Being a medical specialist, it's likely that the various component memes of neurasthenia were prominent features of his attractor landscapes. They coalesced and he got the illness. He was in any case not the happiest of men, it seems, for his father and two brothers were ministers of the church. He would have followed them himself, but for his belief in Darwinism. Like Maria, though for different reasons, he was probably under a good deal of psychological stress and maybe this contributed to what happened.

Beard wrote about his 'discovery,' first in 1869 and then at greater length in 1880. The condition proved wildly popular and, quite soon, lots of people had it. It spread into popular culture and helped to set the tone which produced the weary aesthete of the 1890s. The environmental cause that he himself blamed was the stress and pressure of late nineteenth century life, especially the tight time schedules. Other theorists soon came up with other suggestions. In his own way, Beard was a bit like the eighteenth century obstetricians who used to go around spreading puerperal fever. Memeplexes can indeed behave like germs or other physical causes of illness. We nevertheless have to forgive him because it is so often difficult to distinguish the two, and many illnesses *are* due to bugs and the like. Certainly the medical science of his day could not reliably distinguish the two types of cause. He got it wrong, but if he hadn't he would be remembered with an honourable place, instead of an iffy one, in the medical textbooks. Moreover, he was not entirely alone. Another American, Jacob da Costa, discovered a closely related condition called 'effort syndrome' at around the same time. Clearly there was a high probability at the time that these memes would get together, in some combination or other and in someone or other. Beard was mainly unlucky in being the 'someone.'

What conclusions can we draw from these two case histories? 'Memes can be nasty and damage your health' is an obvious one, but doesn't get us very far. Is it possible to do better? I think so. The stories can tell us something about the landscapes within which memeplexes form, and also hint at the proper role of memes in our mental ecologies, when they are not causing illness and mayhem.

The very word 'memeplex' has acquired sinister overtones due to stories like the ones I've recounted here. It might have been wiser, you may think, if I had refrained from helping to spread alarm and despondency. My problem was that the perception already exists. If I had kept quiet about the potential nastiness of memes, you might reasonably have thought I didn't know what I was talking about. And I'm in far more distinguished company than I deserve, for Richard Dawkins told such tales and also appeared at times to be trying to attach similarly unpleasant overtones of meaning to genes, with his picture of them using us solely to promote their selfish ends. But genes don't have purposes, as Dawkins knows better than most of us. We have aims, goals, intentions and all that. Genes simply exist, and they do only what the 'blind watchmaker' makes them do. The 'selfish gene' was an attention grabbing image, but quite inaccurate. Luckily, the promise of gene therapies and the like seems to have outweighed public perceptions of genetic nastiness. They are now seen as mostly benign and lovely, devoted over millennia of toil to providing us with the best bodies possible — and even if they do go a bit haywire sometimes, that's only to be expected. And the same, I suggest, applies to memes. First, let's think a bit about the landscapes in which memeplexes form.

Individual memes exist as dimples, large or small, in top of the hierarchy attractor landscapes, but they don't get actualized there. These top landscapes are the product of social dynamics and you never get to see a 'social dynamic' as such, you only get to see people shaking hands, holding meetings or whatever. Memes feed back down the hierarchy to individual brain dynamics. That's where they manifest — in concepts that people experience. If lots of people share a particular concept and endorse it for some reason, it goes back up the hierarchy and the 'meme dimple' gets a bit deeper. The dimple probably also extends its basin of attraction, because each person is likely to experience the concept in a slightly different way. Therefore what's fed back up the hierarchy is likely to be broader, so to speak, than what went down.

How do memes get to coalesce so as to form memeplexes? Dawkins implied that it's down to happenings at the level of social dynamics, selecting the best reproducers in a pseudo-Darwinian evolutionary process. Which probably tells us something about how particular memes or memeplexes win out in competition with others. It doesn't tell us how memeplexes form in the first place (nor does it tell us how memes form, but we'll leave that issue aside for the moment till we get to the section on creativity). It's hard to imagine that any inherent properties of the state space of social dynamics alone could re-arrange its landscapes so as to connect formerly separate features. These landscapes, however, emerge from reciprocal interactions with those in individual brains. It seems a lot more likely, as my two stories suggested, that individuals provide the original source of innovation, which then may (or may not) get taken up and amplified by the social dynamic. The problem, therefore, is to do with how individuals connect formerly separate experiences. Looks simple! That's something we all do every day. We're constantly making connections of one sort or another.

Actually not quite so simple because it introduces a whole new topic — that of emotion or 'feeling.' It's a hot topic in neuropsychological circles at present, and still at the stage of generating nearly as much heat as light (some would say more heat than light!). There's not even that much agreement over what emotion is, though everyone thinks it is among the most basic functions of mentality. Some produce lists of 'core' emotions, supposedly hardwired into brain stem circuitry; fear, anger, disgust, love and so forth, from which all the rest derive through processes of refinement and elaboration. Others say there are really only two basic ones, feeling good or feeling bad. There are so many puzzles. Is 'mood' the same as emotion? Are hunger and lust aspects of one emotion or entirely separate? What about higher emotions, such as aesthetic appreciation and the like? Aren't they really just part of cognition in general? There's evidence to support of all these disparate points of view.

Nevertheless there is common ground. Some emotions are hardwired into the brain, especially its evolutionarily more primitive parts. You react automatically with disgust to the sight of vomit; fear or rage to a fist being waved in your face; protectiveness and warmth to creatures with big eyes and little bodies. People are beginning to pinpoint the precise circuitry involved. There's also increasing agreement that emotion, in some form or other, is inherent in all experience; the entirely dispassionate, supreme intelligence of so many old science fiction stories is fantasy only. The function of emotion, thought of in terms of attractor dynamics, is to produce seismic shifts in whole landscapes. As Hollywood used to tell us so often, when you're in love the whole world looks different. And the perceived difference is a consequence of a huge shift in your internal dynamics. The dimple in your landscape that used to represent Mary Smith, that girl in the office, suddenly becomes a vast feature, encroaching on all sorts of formerly separate landscapes, ranging from 'the sort of shirt I like to buy' to 'what I'm going to do with my life.' Lesser emotions naturally have less extensive effects, but they act in the same sort of way. They produce widespread upheaval in pre-existing landscapes, with consequences that are generally unpredictable by us because the reformed landscape is an emergent property of an altered dynamic — and we're unable (yet) to say anything much about what will emerge from any particular dynamic. Nature, however, can in a sense 'predict' the likely outcomes of upheavals, since emotions have been honed by evolution to aid our survival.

It's thus likely that 'Maria's' distress and anxiety, and George Beard's existential angst, were what enabled them to come up with their new memeplexes. But sadly it's not quite as simple as 'emotion causes shifts in attractor landscapes.' For emotions themselves are features of their own landscapes and thus emergent properties of everything else that is going on — including, one has to suppose, what's happening in the landscapes they affect. The stories show that emotional attractors were part of the memeplex, as well as a cause of its formation. 'Feeling better' is an attractor essential to the success of Tarantism. 'Feeling weary' is another attractor, hardwired into our brains. Neurasthenia could not have existed without it. Indeed the condition can be defined as a pathological enlargement of the attractor basin of weariness, enabling the experience to arise in all sorts of inappropriate circumstances.

The fact that emotional attractors are components of memeplexes, as experienced by individuals, makes the attractor dynamics of 'plexes' in people look a bit more complicated than before, but does not present any problem of principle. There is such a problem, one might think, when it comes to trying to picture the top of the hierarchy social landscape. Emotions are essentially about meaning, whereas we've defined memes as being information packets. In any case it's impossible to imagine a social dynamic having feelings. The cop out is that we also regarded memes as being those particular packages of information that are experienced as meaningful when activated in a person (see the beginning of this chapter). This leaves an open question, though. How can meaning be retained in a system (social dynamics) in which it is not obvious that it can mean anything? Paradox!

Or maybe not such a paradox. After all, we regard a good book as highly meaningful, even though information about the pattern of ink on its pages has no intrinsic meaning whatsoever. It all comes from what we project into it. And of course meaning often does disappear in the social dynamic. The experience of mesmerists for instance, which was charged with emotion and absolutely central to their lives, is almost wholly lost to us nowadays. The social dynamic deals with meaning somehow, and does its thing by feeding the outcome back down into our experience. Outcomes may include enhancement or loss of meanings that we originally put into it.

The difficulties in thinking about information/meaning relationships aren't quite as acute for us as they would be for most people nowadays. We are defining information as Bateson's 'a difference that makes a difference,' not as Claude Shannon's 'bit,' which is a far more abstract concept and one that explicitly excludes any concept of meaning. 'Information' in the social dynamic thus refers, according to our definition, to concrete things like particular speech acts or printed words or telephone calls. And it is possible to imagine, at least in general terms, how these could both retain meaning acquired from us as individuals and transmit it back to us. Mirror neurons help with some aspects of the process, but many other mechanisms are likely to be involved.

The main point I want to make is that emotion is both essential to the formation of memeplexes and is part of their constitution — not only as we experience them but also in their abstract representation as features in attractor landscapes formed by social dynamics. Let's move on now to look at the proper roles of memes and 'plexes,' instead of the pathological ones.

You can get some idea of what memes ought to be doing from seeing what goes wrong when a meme turns rogue, just as watching a gang of rowdies in the high street can afford insights into how youth should behave. 'Good medicine is nasty' and 'torture produces truth-telling' are both ideas that occasionally *are* true. The problem with them, moral concerns aside, is due to the fact that they are more often untrue. They are false guides. And surely that's the proper function of memes: to represent aspects of truth to us. I was going to write: 'They put concepts into our minds that correspond with reality and so help us to cope satisfactorily with our worlds.' But that would not have been quite correct. Memes don't so much put concepts into our minds as contribute to creating our minds in the process of expressing themselves.

The same applies to 'plexes.' Their job is to build into us useful skills and ways of thinking, to afford us perspectives on our world that enhance our ability to perceive what's going on and achieve our goals — our *own* goals, note, not those of any 'selfish meme.'

As the function of genes is to provide us with healthy bodies able to read books, play football, fight off infections, raise families, and all the other things we want, so the function of memes is to provide us with similarly healthy minds. However the analogy can't be pushed too far, for it breaks down when it comes to what 'us' means. 'We' are not our bodies in the same sense that we are our minds. I'm walking a philosophical tightrope here because mind/brain identity theorists will be saying: 'Nonsense! Of course we are. Our minds are nothing more than our brains in action, and our brains are part of our bodies.' I don't want to deny that (or indeed to affirm it). However, I think the most diehard identity theorist would agree that there's more to our minds than those aspects influenced solely by genes. There's so much that comes from our families, education, societies and so forth; in other words from ambient memes and 'plexes.' They become part of us in a far more direct and intimate sense than genes, which play their principal part many steps further down the hierarchy leading to the creation of our minds.

What this boils down to is an inherent difficulty in distinguishing our own intentions from intentions attributable to memes that we harbour, for there is a very valid sense in which we *are* our component memes. And if we harbour a 'rogue' ... well, who's to say that it is not just as much 'us' as our non-rogue components? It's quite difficult for a twenty-first century person like me to respond to this question, for we've all been imbued with some of the spirit (the 'plexes') of post-modernism, deconstructionism and the like. We find it quite natural to suppose that, when something is deconstructed, all sorts of semi-autonomous nasties may emerge from the woodwork. And this applies to people, we tend to think, as much or more than to literature, political movements or whatever.

Jung would have found it easier to respond. 'You cannot decide what is alien to their nature for other people' he would have said, rather sternly I guess. 'You can only learn how to decide for yourself, as you achieve individuation.' His answer would have been true because 'individuation' can be viewed as a process of bringing all the memes that we host, especially the ones with a large emotional content, into some sort of balance and harmony. Any that are left out in the cold when the process is complete are 'not me.' Trouble is, few of us manage to complete the process. So we continue to harbour all sorts of ideas and attitudes that can shock if they emerge into the light of day. 'Surely that wasn't *really* me?' we say about them when they raise their ugly heads. Sadly, it usually is 'me' and the best we can hope for is that one day it won't be.

9. Interlude

We've now assembled a pretty good instrument for examining some problems to do with mentality. It's not very sophisticated; you can regard it as being around the early microscope level, and nowhere near the mass spectrograph equivalent that might be ideal. All the same, it does allow us to access and throw light on matters that were formerly hidden away. I want to spend a little time thinking about what it can and can't be expected to achieve. Then we can move on to playing with it in subsequent chapters.

What does the instrument itself look like? Its framework consists in the hierarchy of interdependent attractor landscapes, stretching all the way from the genetic one at base to social dynamics at the top. The lens and focussing devices are the attractors — the memories, memes and archetypes — that exist in these landscapes. You can tell straight away from this that it's going to be useful for looking at the content of consciousness, but is not going to be any good when it comes to thinking about why consciousness should exist in the first place. Perhaps I should explain why not.

The stream of consciousness is mostly to do with higher-level attractor dynamics in brains. What goes on at 'Level 1' is mainly unconscious, though bits of it can sometimes tip over into consciousness if they happen to make a particularly large contribution to the emerging higher level dynamics. Above and below these brain levels everything is always unconscious — so most of us think anyway. A genetic landscape isn't consciously aware, nor is a social dynamic. So it is not attractor dynamics *per se* that make for consciousness; it's attractor dynamics plus some other unknown ingredient that enters into a fairly small proportion of the neural dynamics. My own belief is that the extra ingredient has to do with ability to report on the dynamics. Brains can sometimes directly report to themselves on themselves. The other systems either can't do so, or are limited at best to indirect report. However, that's personal belief only. No one knows for sure what the extra ingredient is.

There are actually still a few people around who might be willing to argue that attractor dynamics alone is sufficient for consciousness, with no extra ingredient needed. Their position is best explained in terms of a famous thought experiment, known as the 'Chinese room,' which was proposed by philosopher John Searle over twenty years ago and is still provoking debate. The set-up that he imagined consists of a sealed room containing a man who speaks no Chinese plus a lot of very detailed instruction books, written in English, telling him rules for altering all possible sets of Chinese ideograms into other ideograms that will make sense to a Chinese speaker. Chinese people post questions into the room, written in their own language. The man inside looks up his books and, following the instructions he finds, writes new ideograms and posts them back out. To the Chinese outside, it looks as though the room contains someone who understands their language and is, presumably, conscious. In fact, however, the man inside understood nothing, other than the instructions on how to change one set of gibberish (as far as he was concerned) into another.

Searle told this story in order to demonstrate that grammar is not the same as semantics, nor is information flow the same as understanding or, by extension, consciousness. However some commentators argued vehemently that the whole set up, room, man and instruction books, *was* conscious. They often referred to Leibniz who, writing in the seventeenth century, had pointed out that if you could look inside the brain of a man, all you would see would be the equivalent of the gears, shafts and so on, visible if you look inside a windmill. If a conscious individual can somehow be composed from Leibnizian 'gears of the mind,' they said, why not from the instruction books and other parts of the Chinese room? Those willing to argue this way could well suggest that a social dynamic can be conscious even though we inevitably have trouble recognizing this.

The issues have been thrashed out at length, some would say *ad nauseam,* and there's no longer much general sympathy for the 'Leibniz' argument. The main underlying problem is that it applies a reductionist approach to what is probably an emergent, holistic phenomenon. You could no more expect to see conscious mind by looking at 'gears in the brain' than you could expect to see liquidity by looking at individual water molecules. And of course there's no reason to expect that liquidity could emerge from an assembly of *models* of water molecules; equally there's no reason to suppose that mind could emerge from something as unbrain-like as instruction books and so forth.

Indeed most of us think that the 'Chinese room' is very iffy indeed as a thought experiment, even in relation to its original purpose of showing that grammar's not the same as semantics. These experiments usually have to be plausible in the sense that they have to be realistic in principle. But there's no realistic way, not even in principle, that the instruction books and man could achieve what they were supposed to achieve (that is, provision of sensible replies to any possible questions within a reasonable period, such as the lifetime of the questioners). It's remarkable that such a flawed thought experiment should nevertheless have provoked so much constructive debate about mentality and consciousness.

Or perhaps not so remarkable. If you try to draw conclusions directly from a duff thought experiment, you go astray. On the other hand, examining why the experiment is duff can be very useful. Einstein allegedly came up with special relativity as a result of mulling over why his adolescent fantasy (thought experiment) of keeping up with a beam of light was unrealistic. Wittgenstein built a whole career on constructing daft arguments and then publicly demolishing them to reach a conclusion. Problem was his followers didn't always appear to realize which stage he was at, partly because he moved at such a snail's pace from one to the other. They sometimes mistook the 'daft' stage for the conclusion. Searle is in good company, therefore. His idea was a failure as a proof of principle, but a triumphant failure nonetheless.

All in all, I think it really is pretty safe to assume that our 'microscope' is not going to tell us much about the origins of consciousness as such. Daniel Dennett tried hard at one time to use some of its components to 'explain' consciousness. He's not into attractor dynamics, but does have a lot to say about memes. The conclusion reached in his earlier work *Consciousness Explained* is that it's illusion-like, but that's hardly surprising because he was using the wrong instrument for the job. If you use a lens to try to find an electrical field, say, you're bound to end up thinking the field isn't really there. He now agrees that consciousness itself is 'real,' though its content is illusory in the sense that it's a *post hoc* construction built from a proportion of the information available to the brain — which is fair enough. And he also showed that using memes can throw a lot of light on the contents of consciousness. Maybe our more complete instrument can show us more still. After all, we've got a structure that links everything that goes into the making of mind, from genes to society. And we can incorporate emotion into the structure; memes considered on their own can appear pretty cold fish, but minds are not like that. Given these advantages, we should be able to see a bit further than Dennett.

The big question is where best to look. The most obvious practical applications of our new perspectives are to medicine. We've already seen how neurasthenia was a mainly social construct with added components from lower down the hierarchy, but was nevertheless considered to be a physical illness (the assertion that it was 'non-mental,' perhaps taken to also include 'non-social,' was one of the memes that went into its make-up). There are lots of conditions to which much the same considerations could apply. For example eating disorders whose popularity seems wax and wane so markedly, or the 'total allergy to the twentieth century' that was popular in the 1980s and now seems to have vanished, or 'agoraphobia' which used to be the single most common condition met in psychiatric out-patient clinics forty years ago and was often of crippling severity — it seems both rarer and milder these days. One may also wonder about current conditions such as Gulf War syndrome, or the epidemic of 'attention deficit hyperactivity disorder,' which has resulted in huge numbers of children being prescribed a drug (Ritalin) that could get you arrested if you were to take it in a nightclub.

I'm not claiming that all these conditions definitely are or were closely analogous to neurasthenia. Merely that it's a possibility and the method of looking for contributory memes could help to clarify which, if any, are in fact like neurasthenia. It's a method which would help to remove a lot of contemporary medical confusion, if people came to accept it. The main conceptual hurdle in the way of acceptance, which few manage to clear, is the perceived trichotomy between 'physical,' 'mental' and 'social.' This causes all sorts of confusion in medical personnel as well as the public when it comes to thinking about the causes of disease. And, as our 'hierarchy' shows, it is largely illusory.

Identifying contributory memes would have implications for treatment and prevention of illness, not just for mitigating confusion. Countering the 'rest is best' meme was important in the treatment of neurasthenics. Loss of belief in neurasthenia's 'non-mental' origin seems to have helped the whole thing to evaporate. Medical historians could take on a whole new role! They could look at the history of ideas contributing to current disease and become 'historico-diagnosticians.'

But that's for the future and the issues involved are likely to prove quite technical, as well as boring for non-specialists. There are probably comparable issues in all sorts of other specialist fields — politics and economics spring to mind. However, since this is only a preliminary survey of a huge new field, it's best look at some of the more general topics in the rest of this book. The specialists will have their say soon enough!

10. Manipulating Memes

According to that too well-worn piece of wisdom, death is to our society as sex was to the Victorians — unmentionable. In fact there are many 'unmentionables'; topics that are ubiquitous yet disquieting in some way that threatens the very foundations of our self-image or what we think our societies ought to be like. One of these is advertising. We turn away from its ubiquity as respectable Victorians pretended that sex had nothing to do with them personally. When it is forced on our attention, we attempt to hedge it about with skirts analogous to those that concealed the legs of every pretty miss and her piano.

Everyone does it. When I was in North America in the 1980s, each time I met some academic in his (it was usually 'he' then) office, he would make a few introductory remarks, then say, 'And this is my résumé,' handing over a bulky document. It would be beautifully printed and laid out, and would list every last review and letter that the academic had ever published, sometimes going back to high school days. Another section would detail minutiae of work, committee membership, voluntary activities and so on. It was a habit that bemused me at first, for British CVs at the time tended to be one or two page affairs listing nothing much more than marital status, degrees, employment and any major publications. The American blockbusters looked like plain bad taste to me and far too pushy. My gut instinct was that anyone who could make such a song and dance about routine committee work and the like must be quite sad. Yet I should not have felt superior, since we Brits advertised too, with dress codes and pseudo-modest mentions of 'the paper I'm working on' and that sort of thing. At least the Americans were being up-front and direct. You knew where you were with them and what they were trying to sell.

Drug companies spend huge sums on advertising. The friendly local drug rep is quite likely to be earning more than the doctor he or she is visiting. Doctors on the receiving end of all this, especially those benefiting from drug company help to go to conferences in nice Caribbean settings for instance, frequently aver: 'Oh that sort of thing doesn't influence *my* prescribing habits. I go on the evidence and what I read in my professional journals.' Tell that to the marines! There's no way drug companies would spend so much of their profits on advertising if it didn't pay. It's true they are caught in an arms race with one another, which fuels a lot of their spending. Nevertheless the advertising is effective. Luckily, even though many doctors are still in denial, as we all tend to be about advertising in general, others have caught on. Some of the covert and more dubious ploys used by drug companies to put their products in the best possible light are beginning to attract attention.

Advertising is, of course, the art of trying to promote the spread of selected memes ('This product is best,' 'I'm a wonderful, hard-working person,' and so on) and minimize the impact of others ('There's a lot to be said for brand B,' and so on). We'll be taking a closer look at it later in this chapter. First, let's consider one of the more traditional approaches to meme manipulation, which still has many practitioners today — the stick, as opposed to advertising which is more of a carrot.

Controlling memes by force

Attempts to control memes by physical coercion have a history which is probably as long as that of humanity itself. The first instinct of Millenarians down the ages, for instance, has often been to burn books, and sometimes people, suspected of harbouring rival memes. There has long been a hankering after gentler methods. Here, for instance is the great French author Rabelais, who was himself a priest, writing in about 1530 on the subject:

> When Ponocrates saw Gargantua's vicious manner of living, he decided to educate him differently. But first he bore with him, knowing that Nature cannot without great violence endure sudden changes. Therefore, to make a better beginning of his task, he entreated a learned physician of that time, Theodore by name, to consider if it would be possible to set Gargantua on a better road. Theodore purged the youth in due form with black hellebore, and with this drug cured his brain of its corrupt and perverse habits. By this means also Ponocrates made him forget all that he had learned from his old tutors.

Alas, hellebore, even when black, does not really have such an effect and people have always felt it necessary to fall back on crude compulsion. The contemporary account of what happened to a thirteenth century French stonecutter was apparently fairly routine for the time. He believed that this world is the only one that exists, and incautiously said so in public: no heaven or hell in other words. He was imprisoned for sixteen months and thereafter was supposed to wear crosses, which would have been heavy, cumbersome affairs. He did not perform this latter penance and was arrested again. He spent five and a half more years in prison, and still had to wear the wretched crosses after release. Whether this converted him to belief in the next world is not recorded. It must have greatly reduced the attractions of this one, so perhaps leaving more room for the desired ideas. Nowadays we might expect bitterness and resentment to fill the vacated space rather than hopes of heaven, but possibly inquisitors knew the likely reactions of their own contemporaries better than we might suppose. Also, modern inquisitors cause severe physical and psychological distress to their victims in the interests of increasing suggestibility by physiological means; the 'brainwashing' so popular in communist regimes of fifty years ago. Maybe thirteenth century inquisitors had the same aim.

Punishments were not always particularly severe, however. One John Florence for instance, who said in 1424 that the pope was rotten, had only to agree to keep his opinion to himself and be 'disciplined' in a cathedral before the assembled congregation on three Sundays in succession. The disciplining in question seems to have taken the form of what a modern employer would call 'counseling,' rather than anything very draconian. Presumably the authorities sympathized to an extent with John's opinion.

Henricus Cornelius Agrippa was a sixteenth century author who wrote books on occult subjects and thus had a vested interest in keeping a wary eye on the Inquisition. He needed to take a realistic view of their opinions and methods in the interests of his own safety. Guiding principles behind actions of the Inquisition of his times, he said, were that:

> They demand nothing else as the mark of faith than that the offender believe in the Church of Rome. If the offender professes this belief they say straightaway that the Church of Rome condemns something he has said as either heretical, sinful, offensive to pious ears or subversive of ecclesiastical authority. And immediately they compel him to recant and revoke what he said. But [if the offender argues his case] they interrupt him with great noise and verbal abuse, saying that he was not [addressing] a conference ... but [was]

> before a tribunal of judges. ... they show him faggots and fire, saying that in the case of heretics they are not allowed to contend with arguments and appeals to Scripture but only with faggots and fire.

In other words: swallow the whole of what we say or face burning; a direct and simple method of persuasion. Agrippa noted too that inquisitors might not be honest:

> ... when I was in Milan, several Inquisitors tormented a number of honest matrons ... and privately milked very large sums from the poor frightened women, till eventually their cheating was discovered and they were severely treated by the nobility, only just escaping fire and the sword.

Being an inquisitor in the service of Rome was evidently just as corrupting as acting for the Gestapo or a communist party. But how did shepherds of the faith turn into wolves preying upon good, as well as straying sheep?

At the beginning of the thirteenth century, the Catholic Church faced its most severe challenge since the Muslim invasions of Europe. A religion had established itself in the decaying, courtly society of the south of France around Languedoc that was an old enemy in a new guise. It was a form of dualism (equal powers of good and evil) called Catharism. Originally due to the Persian prophet Zoroaster, who was an approximate contemporary of the Buddha, dualism had been cast in a more extreme form by Marcion in the second century AD and by the prophet Mani (from whom derives the term 'Manicheism') in the third. St Augustine himself had been a Manichee until he saw the truth and converted to Catholicism. It re-surfaced in eastern Europe as the Bogomil heresy in the tenth century and got a major foothold in western Europe, especially Languedoc, in the twelfth, probably having been brought there by crusaders returning from the east. In small areas, Cathars comprised up to a third of the population, and their missionaries were active throughout Europe. There's a suggestion that St Francis of Assisi's father, a prosperous Italian merchant, may have been a Cathar sympathizer.

As believers held that the world is irredeemably evil and should be quitted as soon as possible, while introducing new souls into it by having babies is a great sin, it might be thought that the religion would rapidly have self-destructed. However only full initiates, called *perfecti,* were required to act on these principles and most sympathizers delayed initiation until they felt that death was near. Cathars therefore reproduced as vigorously, if not perhaps as merrily, as anyone else. Moreover *perfecti* were very impressive by any standards but especially by twelfth or thirteenth century ones, being austere, chaste, honest and indifferent to worldly vicissitudes. They contrasted with many Catholic clergy of the time, and the reforming, mendicant orders which had their origins in that period (Franciscans and Dominicans) consciously imitated some Cathar ways.

Dualist churches which came to light in northern Europe in the mid twelfth century, particularly in Cologne and Liège, were suppressed with violence. Burnings and lynchings were necessary over a period about eighty years before the orthodox could feel secure. The reaction in the South to the threat was initially much milder, following the spirit of advice given by a bishop of Liège in the previous century, who had written:'We are not entitled to deprive heretics of the life which God gave them simply because we believe them to be in the clutches of Satan. ... Those who are our enemies on earth may, by the grace of God, be our superiors in heaven.'

The first reaction of authorities in the South was thus not to burn heretics (though a few were lynched by Catholic mobs) but to convert them. St Bernard preached at them. Papal legates in gorgeous clothes were sent to argue them out of their errors. Noting the lack of success of these measures, the young St Dominic got permission from the pope to meet *perfecti* on their own ground and went about the villages barefoot and in coarse cloth to point out the truth. He, too, achieved little. The next step was to blame the local Catholic hierarchy and feudal lords of Languedoc. Commissioners were sent by the pope to replace where necessary bishops and to put pressure on local rulers, especially the Count of Toulouse who was suspected of being soft on Cathars.

Still no success. Papal patience was in any case wearing thin when a legate was murdered by someone from the household of the Count of Toulouse. This was the final straw. Grim Northerners were called in to apply the methods so successfully pioneered in Liege and elsewhere. The principal cleric was a man named Arnald-Amaury, abbot of Cîteaux, an enthusiastic destroyer of heretics. He is best remembered for his immortal remark after a Cathar stronghold had been captured and a decision was needed about who should be spared. He ruled: 'Kill them all. God will know his own!'

The most committed lay leader to emerge was Simon de Montfort (father of the Simon de Montfort who tried to usurp the English throne), an equally ruthless man whose efforts were spurred on not only by 'righteousness' but also by greed for land. Towns were sacked, countrysides devastated, heretics burned in their hundreds, though probably not thousands because, while *perfecti* often went to the fire with enthusiasm, Cathar sympathizers usually recanted and were often spared.

Eventually, after the death of Simon de Montfort, the affair finally ended when the King of France himself was tempted into annexing Languedoc. Catharism was then eliminated by patient persecution, which mainly involved depriving suspects of their livelihoods rather than their lives. A great lesson had been learned by the Church: violence works, persuasion doesn't. There was one qualification; namely that the violence must be persistently and consistently applied. A bureaucratic organization, the Inquisition, developed to ensure that these maxims were followed.

Application of the lesson was tempered in most countries most of the time by legal and sometimes by charitable considerations. It found fullest expression in Spain where, in the late fifteenth century, Ferdinand and Isabella (actually queen Isabella was mainly responsible; she was a lot more pious than her husband) unleashed the inquisitor Tomas de Torquemada. Profoundly mistrustful of Jews and converts from Islam, this man was in the mould of Himmler or Beria except that he was far more personally austere. He organized a culture of terror and spurious confession. Thanks to the tradition which he established, some towns burned people every month for periods of several years in succession. Though its practice mellowed with time, the spirit of the Inquisition in Spain allegedly did not soften in the three hundred years from establishment until it was suppressed by Napoleon Bonaparte in the early 1800s.

The organization was however, by its own lights, a success. The Holy Office kept Catholicism supreme in Spain over a very long period. The Spanish homeland avoided most upheavals and perils of the Reformation at least partly because of inquisitorial bureaucracy. It was instrumental in preventing the domestic religious wars which plagued much of the rest of Europe. Moreover it killed fewer people in three hundred years than did the Spanish civil war of the late 1930s in three. Maybe, if the Holy Office had still been possessed of its sixteenth century powers, the civil war would not have happened. By the cold calculations of *realpolitik,* its methods can be justified as representing a 'lesser evil.' The Holy Office itself was aware that its primary role was to suppress ideas rather than people, for it usually punished people for what they said, not what they did. Control of memes, not punishment of sinners, was the main aim.

What are we to make of all this? One lesson to be learned is that memes are indeed central to a person's nature. Both the Cathars and their persecutors were people with families, businesses and normal, everyday interests. Yet they were willing to die or kill for the sake of extremely abstract notions that, from our perspective, were not very different from one another. Apart from theories about the nature and role of Christ, the main difference between Cathars and Catholics was that the former regarded the principle of evil as co-equal with God and the latter didn't. Yet this had little or no practical application. If you were a Cathar, you endeavoured to reject the world and Ahriman in order to ascend into the light; if a Catholic, you tried to reject world, flesh and devil in order to be found worthy of heaven. Yet, somehow, these theoretical differences found their way into the core of peoples' self images and justified every sort of sacrifice and beastliness. Clearly memes carry with them not only ideas but also the most intense passions.

It is interesting, too, that the landscapes in which these phenomena had their main existence were those belonging to social, not individual, dynamics. After all it took three generations of persecution to suppress Cathar memes, so they must have been regularly re-invigorated in individuals by a semi-autonomous source further up the hierarchy. The rituals and social structures of Cathar life were no doubt responsible. Memeplexes like those of Alien Abduction, which are mainly features of landscapes in individual brains, are far more ephemeral and unstable. Although they feed up into social dynamics to some extent, they get little return reinforcement from it and therefore lack the staying power shown by Catharism. As it happened, Cathar ideas never entirely disappeared. They surfaced in alchemical and Gnostic systems in subsequent centuries — they had had a home in Gnosticism ever since the Hellenistic period, and were never dislodged from it — but they were unable to regain a sufficiently coherent social dynamic to allow take-off.

This may seem just a roundabout way of putting the commonplace observation that social structures sometimes develop to reinforce particular ways of thinking and behaving in individual people. Nevertheless, it allows one to see why St Dominic's hope of persuading individuals to return to the Catholic fold was always unrealistic. The main generator and sustainer of Cathar ideas lay in Cathar society, not individual Cathars as he supposed, and no amount of preaching at individuals was going to make much difference.

As the Inquisition came to realize, the only solution was to break the society. Had they listened to the eleventh century bishop who preached tolerance they would never have painted themselves into such a corner, but their own social dynamic afforded no room for wishy-washy notions like his. Nothing less than tough love would do, they considered, and of course they knew in their bones that they were right. The sad history of twentieth century persecutions shows that the Inquisition's insights are still compelling in the eyes of many regimes, which is not a cheerful thought especially as the 'love' bit of 'tough love' is usually omitted nowadays. It's time to get back to advertising, starting with a look at the memes behind it.

The power of advertising

The word 'advertising' originally implied nothing more than letting people who might want your services know of your existence. Putting up a signboard or crying one's wares were traditional means of achieving this. The French essayist Montaigne, writing in 1595, advocated setting up central registries of goods and services available in each area. A number of attempts were soon made in both France and England to do this, since it would so obviously be a useful public service.

The French attempt, by a philanthropist named Theophraste Renaudot, was the most thorough. A physician and a *protégé* of Cardinal Richelieu, he was active in the first half of the seventeenth century and is best known for schemes to provide free medical services to the poor, as well as free medical education. The advertising seems to have been a sideline. These activities, including the advertising, were anathema to the fee-dependent physicians of Paris, of course, who accused him of all sorts of political misdemeanours. Luckily the King believed him, not them, but his projects did not survive his death (in 1653). All these early enterprises flopped sooner or later because of financial difficulties or hostility from vested interests, but they did contribute an enduring tradition. The idea of advertising got connected with those of beneficence and utility at a very early stage.

Politicians of course have always used rumour, literature and the arts in general to promote their own interests. Grandiose Assyrian statues with their inscriptions about slaughtering enemies were crude, but presumably impressive to contemporaries. The Romans, on the whole, were more subtle. Their coins and imperial portrait busts still effectively convey to us nowadays the messages that they were originally intended to transmit about nobility, toughness, spirituality, and so on (different messages were broadcast in different periods of the empire). Moving on to a later period, Oliver Cromwell was a particularly effective user of what's been described as the 'ill-famed near sister of advertising — government-controlled political propaganda.'

Marchamont Nedham, that 'inky wretch' as he was called, was an archetypal seventeenth century advertising man, who wrote propaganda for Cromwell among others. Contemporaries got very indignant about him because he seemed to have no principles. According to which way the political wind was blowing and his paymasters were providing, he could be fervently anti-monarchical or an enthusiastic royalist. He was a man who simply promoted whichever ideas brought him advantage at a particular time, without regard to their implications. This would be the whole core of advertising, which is essentially the art of the Trickster, so Jung might have said, were it not for the rather incongruous public service ideas that had been there from the beginning.

It's a core whose manifestations have changed very little over time. Any modern estate agent, for instance, might be quite proud to have written the following eulogy which dates from around 1680:

> In the sweet Up-land of Luisham (that is, Lewisham) near the new water is to be let or sold a House of good Title and Commodities, with two acres of good land, good Fruit, two Fish-Ponds well stored. Enquiries for further satisfaction at Mrs Terries against the Church. (Quoted in Elliott 1962)

A few years later this sort of thing was called a 'puff' and advertising itself was referred to as puffing. The term self-destructed after a time because it came to be associated with the promotion of worthless quack remedies, and also with the disastrous South Sea Bubble (one of the first stock market crashes); circumstances which were, after all, natural consequences of core activities of advertising. A complaint made by Henry Fielding around 1745 was typical of the period. He wrote that the puffer's art consists of 'throwing such a mist over his readers' or hearers' understanding as shall enable him to impose a cloud upon them for a castle.' This was thought undesirable, not only by the public as was natural, but by many puffers also. It entailed deliberate misinformation which didn't fit the public service part of the image. The term 'puffing' disappeared and it was back to 'advertising' with its overtones of social respectability and usefulness.

Nowadays we have hype instead of puffing. Indeed the whole field has become an example of the sort of social dynamic that feeds on itself. It has a lot of influence on dynamics lower down the hierarchy, but often seems to lack dependence on them. There is the same whiff of frenetic, self-serving and vaguely corrupt activity that is so apparent in professional football and other popular sports. It sucks people into its own dynamic, regardless of their needs or prior intentions.

The oft-told tale of VHS versus Betamax provides a good example. When various alternative video recording technologies were new and competing with one another, these were the two frontrunners. Betamax was technically a little better; but we all came to use VHS. The moral of the story for most commentators is that it shows the consequences of 'critical mass.' The two technologies were incompatible; VHS recorders happened to gain a slight edge in terms of sales at one point and a bandwagon effect kicked in. People started to produce VHS tapes in preference to Betamax because they could sell a few more copies, which resulted in the sale of more VHS machines because Betamax could play only an ever diminishing proportion of available tapes, and so on until Betamax vanished. But why VHS gained its edge in the first place is a question less often asked. It was definitely the inferior system in several ways that mattered to consumers. Tapes were bulkier and picture quality poorer. VHS should have been dead in the water. The win has to be attributed to advertising, in this case entirely divorced from its roots in social utility.

A gloom-monger could use what goes on in modern commercial advertising to suggest that memes have spun out of control in that they now pursue their own dynamic in their own social sphere, regardless of our best interests. We don't manipulate memes, he might justifiably claim, they manipulate us. If he then moved on to talk about political spin ... well, you'd be there all day! However, this has always been true of every society and it's best to look on the bright side. Just as sport is said, with some truth, to be a substitute for war, so advertising can be regarded as a substitute for inquisition-like activities.

Although the memeplex 'advertising' can be regarded as beneficial insofar as it's the lesser of two evils, there is something a bit worrying about the way it seems so totally centred nowadays on its own thing. How can we get memes to work directly for our benefit? How can we know that we are succeeding, given that, in a sense, memes are us?

The idea of truth

There's a very special meme that can act as a universal acid in relation to all the others. Daniel Dennett pictured Darwinism as an acid like that; one able to dissolve prejudices and false conceptions in a whole range of fields. The idea of 'truth,' however, is even more fundamental and widely applicable than that of Darwinism. It has always been special. Other memes compete to get it on their side as it were — 'This belief is *true,*' they shout. They are wise to do so. Any meme that is generally seen to be untrue soon dissolves in the acid, and disappears from the social dynamic. Unfortunately, however, the truth meme is not altogether simple and is therefore open to both misrepresentation and attack.

The attacks used to be straightforward. 'Truth is what we say it is,' proclaimed communist parties the world over. Now the main threat is from postmodernism, that most insidious of philosophies, which started off as one of the intellectual games French philosophers like to play. These do little harm in their own country where everyone knows what it's about and they are treated in the appropriate way — much the same as star football players or leaders in the *tour de France.* Unfortunately, however, this particular game got taken up and taken seriously by others. 'All truth is a social construct and all is relative' enthusiasts proclaimed, which was true enough in its way. Then they took the next step, following the spirit of *liberté, égalité, fraternité.* 'Therefore all truths are equal,' they concluded. Ooops! No, they are not! Luckily it's an hierarchical world and some truths are definitely more equal than others. What makes them so?

The short answer is either conformity with logic or conformity with observed reality, but it's a bit too short. There are many 'ifs' and 'buts.' Let's take a quick look at some of them, logic first:

This used to be considered straightforward. Aristotle's rules were what mattered. There are three of these:

- everything is what it is: that is, if something is correctly called A, then it is A not B;

- if there are contradictory statements, C and not-C, they cannot both be true;

- every statement C that concerns any aspect of the real world is either true or not true.

The rules may look right, but they aren't. Number one is sort of correct, but is more about semantics than truth as such. Two is definitely not true of the quantum world, while three is correct only if the 'real' world is taken to be the 'observed' world rather than its quantum underpinnings.

If Aristotle won't quite do these days, what about mathematics? Surely that is the essence of logic. In fact, it isn't. Godel's famous incompleteness theorem showed that there are 'true' arithmetical statements which can never be proved arithmetically. Turing's halting theorem showed that some mathematical problems are 'uncomputable.' Then there are branches of maths which are 'true' in the sense of being logically consistent, but are known to depend on your pre-suppositions (for instance, non-Euclidean geometries). One up to the postmodernists, you may be thinking. But note that mathematicians and others used logic to arrive at views about the 'truth' of things that are not directly susceptible to logical proof. Logic does have an essential place in underpinning 'truth,' though it is not as straightforward as often assumed and indeed is not fully understood at present.

Next up is conformity with observed reality. And there we have the whole basis for the success of science, which depends entirely upon generating ideas (new memes) and testing them to see whether they correspond with what exists 'out there.' The ideas are only relatively true, never absolutely true. Absolute truth is a goal like infinity; for ever approachable but never attainable. The philosopher and refugee from communist Hungary, Imre Lakatos, showed that beautifully in his *Conjectures and Refutations.* But he also showed, equally convincingly, that successive explanations can get truer in the sense of getting deeper and more general. That's what science achieves when it's working properly, and it's what the post-Modernists refuse to understand. Science performs a dance with logic, mathematics and observation from which ever increasing degrees of truth emerge.

The field of science, however, is circumscribed. Despite the best efforts of 'political scientists' and the like, the dance works properly only in physical sciences and biology. So we remain on shaky ground whenever we try to use the 'truth' meme elsewhere. Radical skeptics like to claim that it should never be used except when supported by unimpeachable logic or shown to correspond with repeatable observations. They certainly have a point but are maybe too puritanical. After all their ruling, followed strictly, would exclude using the idea of truth in relation to 99% of human experience. Yet we know that most of the memes we harbour must possess some degree of truth. They have co-evolved with us and have thus contributed to our survival. Darwinism, that other 'universal acid,' ensures that what contributes to survival tends to correspond to reality. If our eyes always told us about phosphenes in our retinas rather than approaching traffic, we should not be long for this world. Equally if our ideas did not, on average, reflect aspects of reality, we should soon die out, just like those sad nineteenth century tribal warriors whose medicine and charms could protect them against rifle bullets, they supposed. The truth meme is applicable pretty well everywhere in principle, although the practice is often problematic. Maybe the safest rule of thumb is to remember that the noisier the 'I'm true' content of a 'plex,' the less likely it is to actually *be* true.

'What about beauty?' someone might ask. 'Surely that's universally applicable, just as much as truth.' Indeed it is. The problem is that it's notoriously in the eye of the beholder. There are no criteria of logic or correspondence with 'objective' reality with which to validate it. Beauty is like the model you admire in a glossy magazine. Truth is the nice girl next door who may, if you are very lucky, warm your bed and cook your food and nag you into mending your dissolute ways (Political Correctness is rarely, I suggest, intimate with truth!). She is a far more practical proposition when it comes to helping you deal with other memes. Indeed it's tempting to suppose that the relative success of Western culture is due to its usually having had a bit more time for truth than has been the case with comparable cultures. Others may have been more inclined to pursue different virtues. Nevertheless beauty does have vital roles to play especially in relation to our next topic, that of creativity.

11. Creativity

Brits of my generation often have trouble getting our heads around the topic of creativity, for it got mixed up in our minds with all that 'two cultures' stuff. C.P. Snow famously remarked of his contemporaries that classicists and artists were totally ignorant of science, while most scientists and technologists knew little of art. From being an observation on the sociology of post-war Britain, what he said got transmuted into an alleged truth about the different natures and the incompatibility of the two fields. Artists were creative and emotional, scientists logical and cold, and never the twain shall meet. Mathematicians were even further beyond the pale than scientists — they didn't become at all comprehensible to us until after Mr Spock from the television series 'Star Trek' had appeared on our television screens. It's hard to say why this misinterpretation should have occurred, for Snow himself was a living demonstration of its untruth. He had been a successful scientist before becoming a yet more successful novelist. Even Edward Teller, 'father' of the H-bomb, *aka* Dr Strangelove, archetypal robotic scientist, was known to be a superb piano player. Perhaps they were considered exceptions proving the rule, though one could go on multiplying examples of people whose talents crossed 'cultures' for a very long time.

In fact, as we've seen, memes are never passion-free, and that includes scientific and mathematical ones. Moreover the basic activity in each field is the same. All deal with pattern. Mathematics is the abstract study of pattern; science has to do with elucidating the patterns to be found in nature; art is the extraction of meaningful patterns from the physical world, or the imposition of such patterns on it. 'What about historians?' someone might object. 'They are always banging on these days about how misleading it is to look for patterns in history, whose chief characteristic is that it has no pattern.' Well, allowing that current fashion is correct and that historians are right to assume (I nearly wrote 'pretend'!) that the course of history is random, they nevertheless like to tell stories and these are patterns of a sort. Ban stories, and they would be limited to publishing nothing more than tables of statistics or other lists.

Creativity in whatever field is thus all about creating patterns, initially in the mind, that may later be transferred to words on paper, musical scores, paint on canvas, sculpted objects and so forth. They're not just any old patterns. For mathematicians they must conform to rules of logic, for scientists to the structure of the natural world and for artists they must convey desired meaning. But above all and for all groups, they must appeal to their creators' aesthetic sense, their feeling for beauty. We'll get to that later; first let's look at the process of creation.

Henri Poincaré, he of the 'state space' (Chapter 2), thought quite a bit about the process. He was fond of telling an anecdote about how he had been puzzling for quite a while over a thorny mathematical problem (not about state spaces; it was to do with the existence of things called 'Fuchsian functions.'). Then he put it out of his mind one day and went for a trip on a bus. 'Having reached Coutances,' he wrote, 'we entered an omnibus to go some place or other. At the moment I put my foot on the step, the idea [a solution to his problem] came to me, without anything in my former thoughts seeming to have paved the way for it ...' Nor was this an isolated occurrence. Concerning another problem which had been puzzling him, he recounted: 'Disgusted with my failure, I went to spend a few days at the seaside and thought of something else. One morning ... the idea came to me with just the same characteristics of brevity, suddenness and immediate certainty ...' Crossword addicts will be familiar with the phenomenon. If you can't get a clue, stop thinking about it and it will come to you.

Poincaré believed that there are three stages to the creative process. First preparation, involving identifying some problem and mulling over it. Evidently this stage is driven by emotions of various sorts — love of the subject, curiosity, ambition, whatever. It's usually hard work. Next, illumination which often arrives out of the blue, as happened when he stepped on the bus at Coutances. The chemist August Kekulé (von Stradonitz) famously solved the structure of the benzene molecule in a dream. Sometimes it's semi-conscious as in Mozart's account of how music came to him:

> ... thoughts crowd into my mind as easily as you could wish. Whence and how do they come? I do not know and I have nothing to do with it. ...Once I have my theme, another melody comes, linking itself to the first one, in accordance with the needs of the composition as a whole ...

Poincaré too recounted an example of this sort of thing when 'ideas rose in crowds; I felt them collide until pairs interlocked, so to speak, making a stable combination.' Finally, there is the stage of verification.

Is the idea really as good as intuition suggested? Back to the slog again. The bus-stop intuition, for instance, turned out to be valid but only a step on the way (as was, perhaps, appropriate!). It took Poincaré several more years of effort before he came up with a general solution to his problem.

Clearly the moment of creation is linked with the moment of intuition; the other stages have to do with setting the scene and refining the product. But what are the 'objects' on which creation works? Jacques Hadamard, another French mathematician, found himself at a loose end when he was exiled in New York during World War Two. He was a refugee from Vichy France, in his late fifties, lonely and disoriented in a foreign country. His interest in creativity was, perhaps, a form of occupational therapy. If so, it was nevertheless enormously useful from our point of view, for he was sufficiently well known to be taken seriously by the great and the good when he enquired about their thought processes.

Some people, especially academics in departments of English or Philosophy, argue that thought, including creative thought, is inseparable from language. Scientists have to think in words or mathematical symbols, they suppose. Artists may have their own 'languages,' but there's no reason without words of some sort. It's amazing how this totally mistaken idea has persisted. Hadamard encountered and dismissed it, but it still lives on. He thought it due to word-centred academics generalizing their own personal experience to all — whereas many people, and especially creative ones, simply don't think in the same way as scholars of that sort. He quoted Francis Galton, the great Victorian geneticist and polymath, as saying that he sometimes happened, while engaged in thinking, to catch an accompaniment of *nonsense* words, 'just as the notes to a song might accompany thought.' Galton grumbled that: 'After having arrived at results that are perfectly clear and satisfactory to myself, when I try to express them in language I feel that I must begin by putting myself upon quite another intellectual plane. ... I therefore waste a vast deal of time in seeking for appropriate words and phrases ...'

Hadamard went on to point out that a similar phenomenon (that is, having to translate when trying to express creative thought) seems widespread. A well-known economist (Sidgwick) thought of economic problems in terms of visual images. Some composers, too, have seen their initial conceptions in a visual form. Perhaps these people had synaesthesia, a phenomenon of which Hadamard seems to have been unaware. His prize catch, however, was Albert Einstein, who wrote him a letter describing how he thought:

> The words or the language, as they are written or spoken, do not seem to play any role in my mechanism of thought. The psychical entities which seem to serve as elements in thought are certain signs and more or less clear images ... in my case of visual and some of muscular type. Conventional words or other signs have to be sought for laboriously only in a secondary stage ...

All this can be taken to suggest that the moment of creation involves rearrangements of attractor landscapes, perhaps mostly at Level 1. It will be recalled that most of what goes on at this level is unconscious, and that seems true of creative intuition also. The objects in these landscapes are attractors of all types embodying all aspects of mental life, not language alone. And the objects that enter into processes of creation arc similarly various. What goes on is generally much the same as the processes leading 'Maria' to the 'discovery' of Tarantism or George Beard to that of neurasthenia. Under the pressure of emotion, they experienced a rearrangement of existing landscapes, resulting in the joining up of previously separate features. And most creation, especially artistic creation, equally consists in new conjunctions of pre-existing features. Tracey Emin's 'My Bed,' for instance, was novel only in putting a real unmade bed into the 'wrong' context — that of an art gallery. One has to admire her chutzpah, but the creativity involved was fairly minimal. Much the same can be said of Damien Hirst's pickled animals.

But other artists do appear to have created true novelty; Turner and the French Impressionists, for example, or the Cubists. And this sort of creation is a lot more common in mathematics and science than in the arts. Scientists, however, usually talk about 'discovering' rather than 'creating' their novelties — and so do some mathematicians, interestingly enough. Their discoveries are, in our terms, features of their own attractor landscapes. Presumably they are completely new features, which spontaneously develop somehow from the pre-existing dynamics. The creator then 'discovers' the novelty in his own mind which reflects, if he is a scientist, some feature of the external world. If he is a mathematician, it may or may not correspond with any external reality but is a 'discovery' nonetheless.

Unfortunately it's impossible to take this line of thought any further, much as I'd like to do so. We simply don't know enough (yet) about how attractors 'encode' particular mental features or about the dynamics involved, to say anything useful. So let's leave it and ask instead where beauty comes in; the feeling of aesthetic rightness that is so important to creation in all fields?

A sense of beauty

The smallest church in England, so locals claim, is at Wasdale Head nestling beneath Scafell (pronounced *score*fell) the country's tallest mountain.[1] The church may be small, but the graveyard is big. There's only one hotel and a few scattered farms for miles around. Why so many dead? The mountain provides the answer of course. At 964 metres, it's a mere pimple by world standards. Nevertheless it attracts walkers and climbers in their thousands. Some fall off and a surprising number die of exposure, or used to do so before the days of mobile phones and ever-ready mountain rescue teams. I was much impressed, as a child, by the story of two young teachers who had come to the area for a walking holiday. They started up a neighbouring mountain (Great Gable), clouds came down and they lost their way. They were found next day, huddled up in the lee of a rock and quite dead. 'How silly!' I thought at the time. 'All they had to do was head downhill. The path didn't matter.' But it was hypothermia not stupidity that did for them. Hypothermia is less of a problem these days with vastly improved clothing. Nevertheless, despite all the high tech gear, the number of climbers killed in the Alps each year averages nearly three figures. The toll on Everest increases almost annually as travel companies convey ill-prepared clients from the comfort of their own homes to zones of anoxia and altitude sickness. These people are lured to their doom by a sense of beauty that would have been incomprehensible to most before the end of the eighteenth century.

Back then, mountains were hideous wildernesses, full of discomfort and danger, to be avoided if at all possible. The memes were doing their job, in other words, and giving people a realistic picture of what mountains are in relation to us. Nowadays mountaineers have to discover for themselves what they are like, at the cost of experiencing cold, hunger, bruises, chilblains and frostbite. For what they see beforehand, tutored by the Romantic imagination, is not a mass of barren rock but realms of sublime disorder, reaching up to the impossible purity and light that is to be found far above the mundane world. Ugliness has somehow been transmuted, in the course of a few centuries, into beauty. The same thing happens regularly in art, on shorter time scales. The Impressionists were castigated by the critics at first; now a Monet goes for millions, its value ultimately down to perceived beauty. Picasso's famous 'Guernica,' the central feature of which is an angular caricature of a startled horse's head, appeared ludicrous to most of us at first sight, but now seems, if not beautiful in an everyday sense, at least remarkable in an aesthetically pleasing sort of way. These changes in perception do not always persist however. Schoenberg's music went from cacophony to darling of the in-crowd and back to cacophony in a couple of generations.

Does this mean that beauty is just another attractor in the social dynamic? I suspect that it is actually a quite different sort of beast. I suggest that the structure of our attractor landscapes influences what will possess aesthetic value for us, but this value is a quality of the attractors involved and is not itself a separate attractor. Let me try to explain.

Neuroscientists are beginning to take a professional interest in aesthetics. Vilayanur Ramachandran and Semir Zeki, in particular, have developed theories about it. Ramachandran is director of the Center for Brain and Cognition in San Diego. Zeki is an FRS and co-head of a department of cognitive neurology in London. Both, in other words, are extremely creative people at the top of their profession. They have independently come up with ideas that sound rather different, but in fact share similarities.

Ramachandran's suggestion is that satisfying art can be regarded as caricature, which stimulates particular neurons in the brain more intensely than any natural stimulus. We all have 'feature recognition' neurons in our brains that light up in response to things like particular people, places, types of gesture, facial expressions, colours, line orientations and so on and so forth more or less indefinitely. He argues, semi-seriously, that successful artists are able to stimulate such neurons with products that are free of much of the clutter of the 'real' world, so we experience their activity in a purer and often more intense form. Predictably, art historians have considered this suggestion crude and inadequate but Ramachandran counters that at least it's a start. Many others have agreed with him. His approach opens aesthetics to empirical investigation. In his own words:

> Consider a simple question such as what makes a female (or male) face beautiful? [Francis] Galton showed that an average of many female faces tends to be prettier than any one exemplar since it tends to average or iron out any imperfections. But we would predict that ... a beautiful

> female face will result if you subtract an average female face from an average male and amplify the difference. It makes evolutionary sense that blemishes would be unattractive and would be smeared out by the averaging process (as in Galton's pictures). But, if so, why are *some* blemishes — such as beauty spots — attractive? I recently found that if Cindy Crawford's famous mole (near her left upper lip) is moved to the central midline, for instance, on the forehead or tip of the nose, it looks hideous. It needs to be asymmetrically placed. It's also prettier if placed near a sharp facial feature, for instance, the angle of a lip or near the angle of the eye, but not so much if it's far from a feature: for instance, the middle of the cheek. And lastly if *two* moles are symmetrically placed on either side of her nose just above her upper lip, it again looks bad ... the aesthetic response to the precise positioning of the beauty spot is quite lawful yet we have no inkling of why such a lawful response function exists.

This observation about the mole might not have surprised the Nobel prize-winning astrophysicist Subrahmanyam Chandrasekhar, who liked to write about aesthetics and the role of beauty in creativity. He quoted with approval Francis Bacon's (the sixteenth/ seventeenth century Lord Chancellor and Philosopher) dictum that: 'There is no excellent beauty that hath not some strangeness in the proportion!' — adding that Werner Heisenberg's assertion also applies: namely, 'beauty is the proper conformity of the parts to one another and to the whole.'

Zeki's theory is to do with the process of abstraction. Information reaching the brain is a buzzing turmoil of flux that we have to make sense of somehow. We learn to abstract constant features from the flux; for instance we learn that a friend's face *is* the friend's face regardless of our angle of view, the lighting conditions, whether he's putting on spectacles or drinking a pint of beer, and so on. We not only abstract perceptual constants, but also principles, such as 'love.' These abstracts have a lot in common with Platonic Ideals, says Zeki, the pure forms that exist in the world outside Plato's famous cave (Plato pictured us as inhabitants of a cave, able to see only shadows of the true reality that exists outside, projected upon the walls of the cave). Art is able directly to activate or express these abstractions in a way that mundane reality can rarely or never achieve. Zeki has given a nice example of how the ideal love expressed in Wagner's opera *Tristan and Isolde* not only expressed Wagner's own yearnings, but also was so ideal that the central characters themselves knew it to be unrealizable — hence death was preferable to its unattainability. Zeki explained:

> Abstraction leads to an Idea or concept, but our experience remains that of the particular, and the particular that we experience may not always satisfy the Idea formed in and by our brains. One way of obtaining that satisfaction is to 'download' the Idea formed in the brain into a work of art.

The two theories are similar in that Ramachandran's feature-detecting neurons are physical instantiations of Zeki's 'abstracts' or 'Ideals.' The activity of a neuron that responds only to the sight of a particular friend's face, regardless of ambient lighting and so on (and such neurons are known to exist), is an 'abstract' of the face. The advantage of Ramachandran's version is that ideas for experiments can be derived more easily from it. The advantage of Zeki's approach, on the other hand, is that one can more readily see how it might apply to art in general; Ramachandran's ideas fit most naturally with rather simple visual images, and come across as a bit 'forced' if extended to other fields. In attractor landscape terms, both are saying that art has to do with things that most naturally activate or express some particular attractor. Zeki's 'Ideals' are functionally equivalent to attractors, while Ramachandran's feature detecting neurons can be regarded as the cell or cells which activate when a relevant attractor comes into play. Can one extract a theory of beauty from this as it stands? I suggest it provides a useful basis for a theory, but nothing more.

The feeling of beauty, aesthetic 'rightness' if you will, is not confined to art. It arises equally in relation to the natural world and to purely intellectual constructs. Einstein's field equations of general relativity, for instance, are universally thought by those in the know to be supremely beautiful. I suppose you could say that natural landscapes we think lovely are exaggerations of familiar ones and thus fit Ramachandran's theory, or that they are more like our Ideal landscape than most of those that we come across, which would be OK from Zeki's point of view. But snowflakes seen under a microscope are beautiful too, and were considered so by those who first saw them: people who couldn't possibly have had any Ideal of a snowflake or snowflake feature detector already in their minds. Moreover hexagonal symmetries like those of snowflakes are quite rare in nature, so it's unlikely that they would have had any closely related abstracts or detectors in their heads. Similarly, physicists who first saw Einstein's equations needed little convincing of their value. They could see their aesthetic value straight away, and half-expected all the rest to follow. They had no ideal 'gravitational field equation' in their minds to which Einstein's could approximate. Rather, they had a lot of other mathematics related attractors already there, among which Einstein's new concept could fit.

And there, I believe, lies the key. Beauty is not a property of any single attractor. Rather it is a measure of how well an attractor fits a pre-existing landscape. Does it snuggle in comfortably and make everyone else function better, or does it stick its elbows out and trip people up? If the former it's beautiful; if the latter ugly. The sight of mountains upset the landscapes of pre-Romantic people but enhanced those of Romantics, so the mountains themselves were horrid to the former and sublime to the latter. We often tend to think that beauty is all about pleasant emotions, but of course it isn't. A frisson of terror enhances appreciation of mountains. Francis Bacon's (the twentieth century painter) pictures of people as gobbets of decaying flesh are quite literally disgusting, but are beautiful largely *because* of the disgust. It's not basically a matter of pleasantness, niceness or anything like that; it's a matter of fit. And therein lies a problem; for the mental landscapes of sadists or of child pornographers, for example, are such that they are honest when they say they find their activities 'beautiful.' Beauty, in brief, is not in the eye of the beholder but rather in the landscape of the experiencer.

There are, however, general properties that are often associated with beauty. One is symmetry or Heisenberg's 'proper conformity,' another fractality. Too much symmetry becomes boring, but the patterns in kaleidoscopes for example are often lovely and this quality depends almost entirely on the reflections. The original unreflected pattern, if looked at on its own, is usually of little interest. Similarly, everyone loves a fractal pattern, like frost on a windowpane or a leafless tree seen against the sky or a reproduction of the Mandelbrot set. The Golden Ratio has some claim to be included here as well.[2] It turns up quite often in nature, in things like seashells and flowers, and big claims have been made for its importance in the visual arts. It appears that approximate golden ratios can be identified in paintings and architecture quite frequently, but they don't have to be exact. A roughly one third/two thirds distribution is generally good enough. Maybe these properties are telling us something about the overall structure of our landscapes. Perhaps the landscapes themselves are symmetrical and fractal, thus allowing anything with a similar structure to 'fit in' more readily. Possibly there is something about the golden ratio that helps new features to accommodate themselves to a pre-existing landscape.

But how could we be conscious of beauty? All that I've said so far relates mostly to Level 1 landscapes, which are largely unconscious. However, it will be recalled that Level 2 both emerges from and orders the activity in Level 1 — and is the main home of consciousness. Harmony in level1 will translate into enhanced functioning in Level 2. That's what we're conscious of when we experience beauty. It's like an oil that enhances and heightens the dynamic of our conscious landscapes, enabling them to transcend the disharmonies and scratchiness of the mundane world.

We've now reached the stage of being able to formulate a general definition of creativity. It's the process of re-arranging attractor landscapes in the brain in such a way that Level 2 functioning is enhanced — an enhancement that is experienced as a sense of aesthetic value or beauty. That's a big step forward from a dictionary definition like 'the ability to create,' which is circular. We can relate creativity to a general property of the brain — its attractor landscapes — and have gained some insight into how it occurs, where aesthetics comes in and why different people may have different aesthetics.

The attractor dynamics picture, which looked at first as though it might have been no more than a castle in the clouds itself, has enabled us to describe how it is that we create castles in the clouds along with more useful concepts. Though something of a brain teaser, I think this strongly suggests that the picture is in fact solid and useful. Something that helps towards accounting for its own existence must have a lot going for it! In the next chapter, we shall explore an aspect of the limits of its potential usefulness.

12. Mystical Experience

The topic of this chapter is regarded as meaningless, trivial or delusory by some people, by others as profoundly significant. Can our new 'toolkit' throw light on it? Let's start off with a brief look at the life and ideas of someone who *did* think it enormously important.

Evelyn Underhill (1875–1941) was well known as a writer on religious topics in the first half of the twentieth century, though few remember her now. She was the only child of a successful barrister who had a love of yachting (he was commodore of a cruising club at one stage). Evelyn shared this. She and her husband owned boats for most of their lives, and frequently went sailing. She had some affection for mountains also, though it seems to have been a relatively conventional, not particularly intense sentiment. Her parents took her to Italy via the Alps when she was aged twenty-three, and she wrote: '... we saw Monte Rosa, and the Matterhorn and all the great Alps spread out all around us. It was the most glorious thing you can imagine ...' But she was soon off fossicking enthusiastically among Italian churches and art. Parents even allowed her to spend a fortnight in Florence on her own — an unusual amount of freedom for a Victorian Miss. Relatively late in life, however, she did fall in love with the Norwegian mountains perhaps because visits to them could be combined with sailing.

She married a childhood sweetheart named Hubert, another barrister who, like her parents, had no interest in religion. They had no children. One gets the impression that his main role in her life was that of a 'Mr Fixit.' When she discovered a temporary enthusiasm for bookbinding, for example, he was allowed to sharpen her knives. He saw to the practicalities of sailing, house maintenance and so forth. However, he was an enthusiastic DIYer and allegedly content with his lot.

By the time of the First World War, she was a well-regarded poetess and writer on mysticism. Her book *Mysticism* (1911) in particular attracted a good deal of attention. Her output was prodigious, for she produced in all forty books in thirty-nine years, plus more than 350 articles of various sorts. A main thesis of her earlier writings was that Christ and St Paul were both primarily mystics, albeit ones with slightly differing messages. Evelyn had a wide interest in other mystics, too, including the contemporary who was attracting so much attention in that period — Rabindranath Tagore. Her personal favourite was said to have been the fourteenth century Jan van Ruysbroeck, a predilection that may have had fateful consequences for her later on. Ruysbroeck was, of all medieval mystics, the one who had least time for heresy (though he was himself accused of it for a while). If some of this rubbed off on Evelyn, it is likely to have been partly responsible for her choice of spiritual director.

In common with many other writers on the topic, she developed a classification of stages of mystic progress culminating in experience of union with the divine. Her particular scheme had five such stages. A highly intelligent person, she would probably in time have come to realize that schemes like these have only limited validity — the phenomenology of mysticism is far richer than is often realized and does not always behave as it is 'supposed' to do. Left to her own devices, she might have continued her studies and made valuable contributions to the whole field. However, her development was hijacked around 1921 and sent down a very different, far more conventional path.

This came about as a consequence of her own decision, perhaps influenced by fears about being heretical. She chose as her spiritual director Baron Friedrich von Hugel (1857–1925), a devout Catholic theologian and fellow writer on spiritual matters, who was also a would-be modernizer and advocate of the idea that there is truth in all religions. He was an Austrian aristocrat, but had been living in England since 1873. Von Hugel somehow managed to convert Evelyn into a 1920s version of a 'born-again' Christian, whose life centred on ideas about the historical Christ. She became a frequent visitor of the poor and needy — some of whom, to be fair, do seem to have valued her ministrations — a conductor of 'spiritual retreats,' and an active, indeed hyperactive, figure in High Church Anglicanism. Jung would greatly have disapproved. It's a pity that she never met him. 'You have to find your *own* path to the great mysteries,' he would have told her. 'It's no good following the one trampled by a billion others.'

Be that as it may, she had in fact encountered von Hugel, not Jung, and followed his instruction faithfully for the rest of her life, which was quite short. There was one final conversion before she died. Although not a pacifist during the First World War, she somewhat perversely became one at the start of World War Two but had little time to act on her new principle. Always a wispy, fragile-looking person it was said, she succumbed to asthma and chestiness in her mid-sixties — helped along the way, one is tempted to suppose, by von Hugel's strangulation of her native genius. His intentions were good; to promote the health and welfare of her soul. And he was widely admired by contemporaries for his 'spiritual stature.' But confidence that one knows what is best for other people is usually misplaced.

Why should a bright girl from a busy, prosperous and non-religious background have become interested in mysticism in the first place? Her late adolescent inclinations were towards a then fashionable and largely atheistic Socialism. She described various mystical experiences in letters to von Hugel, but these occurred after she had met him. Moreover she worried, appropriately enough, that they might have been due to 'auto-suggestion' (von Hugel told her they probably weren't). However it seems that earlier in her life she had had spontaneous experiences, the first possibly in 1907, of unity with her surroundings and with God. This sort of thing is not uncommon and is usually felt to be deeply significant and memorable, but doesn't normally have quite such a profound effect as it appears to have had on Evelyn. There must have been other, unknown factors in her temperament to lead to such a remarkable outcome. Nevertheless spontaneous experience triggered her vocation. Let's see what such experiences are like.

Types of religious experience

The Religious Experience Research Centre, now based at the University of Wales, was founded in 1969 by a distinguished biologist, Sir Alister Hardy, who remained its director until 1976. He is probably best known for being one of the first to propose the 'aquatic ape' hyothesis (in 1960). This is the idea that we are apes who adapted to earning a living from food — clams, mussels, maybe shrimps and fish — gathered at sea or lake shores. There's a lot to be said for it. Although it has never become mainstream, it won't go away either and gets revived every decade or so, sometimes by people who imply it's their own idea. Anyhow Hardy had long had an interest in things spiritual, too, and decided they should be researched. The Centre garners accounts of religious experience that people send in; there are over six thousand in its archives. Staff have published a range of statistical and anecdotal accounts based on these and on smallish surveys that they have conducted.

Lots of people, especially when they are children, get occasional feelings of unity with the world about them, often accompanied by perceptions of its extreme beauty. Wordsworth described the experience wonderfully well in some of his poetry. Others go further and feel that somehow they *are* whatever they are viewing — trees, the sky, a city street, the universe as a whole. The whole experience is usually felt as deeply numinous in some way. For many, it involves an awareness of, or a feeling of union with, the Divine. It comes in all degrees of frequency and intensity. As many as 40% of adults, according to surveys, have had it to some degree at some time in their lives, perhaps only once or twice but they don't forget; it is something that seems to get deeply imprinted in their memories. A few people claim to get it often and even to live with the experience for weeks or months on end. Other varieties are less common, but still far from rare and take all sorts of forms.

Paul Marshall, in his recent book *Mystical Encounters with the Natural World,* distinguishes between 'extrovertive' and 'introvertive' experience. The former involves feelings of unity with the world and/or a God immanent in Nature, plus a whole range of other phenomena, especially heightened clarity and intensity of consciousness, feelings of love, beauty and understanding, disturbances of time sense and so on. Introvertive experience, on the other hand, is to do with union with a transcendent God felt to exist apart from the world, with perceptions of intense light and peace, and what is described as 'pure' consciousness — awareness of being aware but of nothing else. However, as he himself allows, the distinction between the two is not clear-cut.

Other dichotomies have been proposed. For instance some mystics describe their experience in terms of overwhelming light and love; some use words about the passionless impenetrability, the darkness of God. Some say that their experience was all about the vast compassion and intense beauty of the Divine. Others claim that any attachment, especially to the self but to everything else also, should be left behind by those wishing to experience God, who can only be described in terms of what He/She is not. In order to reach this understanding of an incomprehensible God, one must first attain a state of dispassionate contemplation, they claim. These two apparently contradictory approaches to, and experience of, mysticism have long been familiar and are sometimes referred to as the *via positiva* and the *via negativa.* Although apparently totally at odds, the two types of experience have been combined by some mystics, who appeared to regard them rather as physicists regard the apparent paradox of the wave/particle duality — as different routes to approaching what is in fact the same Reality.

Perhaps I can best illustrate what may occur with an example from the Research Unit's archives. It's an unusually complex experience, in that it had several stages or phases, but is otherwise unexceptional.

The subject, a man aged 28, was sitting at home reading after wife and children had gone to bed. The television was off. He was allegedly in good health and drug free, though upset by a recent event in his professional life that had left him feeling morally inadequate. Suddenly and for no obvious reason, he became aware that there was a huge presence, personal and beneficent but kind of overpowering, in the room with him. There was nobody to be seen. Words came into his mind somehow (he did not hear them spoken): 'Why don't you look at things the other way, you silly ass?' They had a sort of jocular quality, like a schoolmaster jollying along a recalcitrant pupil.

Then something happened in his mind, the room disappeared and he was before an immense mountain, green and shrouded in parts in silver mist. Between him and the mountain there was a black, impassable gulf. He glanced about and noticed, on his side of the gulf, a golden arc reaching upwards. Looked at more closely, it was composed of innumerable entities all singing perpetual praises to the mountain. More words came: 'The mountain is God the Father.' 'That's the gulf of infinity.' 'There's the arc of created being.' He was totally awestruck and could think of nothing to do other than repeat the Lord's Prayer.

As he started to rehearse the familiar phrases, the words were taken up by 'the arc of created being' and turned into a chant that appeared to him to 'ring throughout eternity,' for he had lost all sense of time. His being seemed to dissolve and follow the chant into a realm of golden light, though he was aware, too, of himself as an entity sited at the base of the arc. Then, quite suddenly, he found himself back in his living room with no idea of how much time had passed. He later deduced that it could have been no more than a few minutes. The memory of it all, he reported, never appeared to fade in the way that memories usually fade.

That's the sort of experience we're talking about. What has science got to say about it? Well, science has on the whole avoided making any firm pronouncements and has mainly focused on the neural states that accompany mystical experience — or generate it as some would claim in private, even if they are often more reticent in published papers. There has also been interest in what predisposes to having these experiences.

A lot of what goes on in the experiences has to do with what are known to be temporal lobe/limbic system functions; the sense of presence; the distortions of time perception; the feelings of ecstasy; even the feelings of ineffability. In line with this, people with temporal lobe epilepsy can have simple ecstatic/mystical experiences associated with their fits; a few have more structured experiences along the lines of Marshall's 'extrovertive' ones. Moreover a proportion of these people develop what is termed 'religiosity' — a preoccupation with big, inchoate and repetitive ideas to do with God or some particular religion. Ever since Canadian neurosurgeon Wilder Penfield's pioneering work (*circa* 1935–65), it's been known that electrically stimulating parts of the temporal lobe can produce experiences with some of the content of mystic ones. This fact is re-discovered every now and again, and gets a few column-inches in the dailies, under headlines like 'Science finds the God-spot' or 'How science has disproved religion.' It seems to happen about every decade or so, rather like rediscovery of the 'aquatic ape' idea.[1]

Building on findings like these, neuroscientist Michael Persinger published a good deal of work, mainly in the 1980s, suggesting that all mystical experience is due to micro-seizures in the temporal lobes. He claims to be able to produce such experiences reliably in experiments using a form of harmless brain stimulation (Trans-cranial Magnetic Stimulation or TMS). Others dispute this, pointing out that the induced experiences are generally simpler and less intense than spontaneous ones. Vilayanur Ramachandran too has developed ideas in this field, in addition to his suggestions about beauty (Chapter 11). He also gives a principal role to seizures, though a rather more sophisticated one than Persinger's as he includes parts played by emotional adjustments following seizure. Indeed most writers on the subject have abandoned Persinger's rather simplistic model. Newburg and d'Aquili for instance, the former a neurobiologist and the latter a psychiatrist/anthropologist, hypothesize that all sorts of brain areas are involved in spiritual experience.

It's more or less impossible to carry out direct experiments to test theories of this sort, because you can't have a spontaneous mystical experience to order. People have turned to meditation in the hope of finding answers because it can sometimes appear to be like, or maybe to induce, mystical experience and *can* be performed in the laboratory. Plenty of studies have been done on meditators, who show all sorts of brain blood flow and electrical activity changes accompanying their practice. Generally, there is increased coherence and sometimes frequency of electrical activity, along with diminution of frontal lobe blood flow. However, details vary a good deal from experiment to experiment. There are known to be differences between the findings in experienced and inexperienced meditators, but otherwise knowledge about the effect of experimental variables on findings seems hazy at present. So far as I know, no-one has observed temporal lobe seizures in meditators.

In fact, it's not at all clear how these experiments relate to mystical experience as such; meditative states are quite often not particularly like extrovertive experience and even the deepest are possibly no more than superficially similar to introvertive experience. There's overlap, but not necessarily identity, between mystical and meditative states. Indeed, mystical experiences overlap to some extent with a whole lot of phenomena, including our old friends the OBEs, NDEs and ayahuasca experiences. This suggests that they are based on some quite fundamental property of the brain that can express itself in all sorts of circumstances, but makes it hard to pin down what the property might be. All one can say for more or less sure is that the temporal lobes of the brain are involved somehow. Another problem with investigation in this area is that it's difficult, or sometimes impossible, to resolve 'chicken and egg' type questions. For example, 'Did the experience cause the brain change I observed, or was it the other way round?' leading on to: 'Is this sort of question actually meaningful?'

As in the case of NDEs, the memorability of mystical experiences suggests that they are special in some sense, and this quality is not necessarily shared by meditative experiences. Unlike most meditative states, mystical ones stick in memory in just the same way as things that are both perceived as hugely important and which come as a shock. The paradigmatic question here of course is: 'Where were you when you learned of President Kennedy's assassination?' Personally, I was in Waterloo Station, London, and heard the news over the station announcement system. My fiancée was with me, I recall, but I've no memory of why we were in the station or where we were going. Most of us, if old enough to remember the event at all, are in the same boat. Mystical experiences are like that. They remain etched in memory long after associated memories have faded.

It is clear, however, that memes are involved in the detail of the phenomenology of many mystic states. In the example I gave earlier, there were lots of memes. 'The mountain of God' has a long cultural history, going back to Moses and Mount Sinai or further; the idea of infinity is a meme, and so is that of the Lord's Prayer (the prayer itself is a 'plex'). The fact that the mountain was green and partly shrouded in silver mist probably derives from something in the personal history of the experiencer. But what about the experience of (partial) dissolution into the song of the 'arc of created being'? It would be pushing the boundaries to call that a meme. It's an experience of a type reported by many meditators, that does not necessarily relate to anything in their cultural histories and often appears to have taken them by surprise. In those respects it resembles the 'black pumas' of ayahuasca visions.

There's a sort of snobbery in mysticism, relating to whose experience is the 'purest.' Who has had the 'deepest,' most immediate or ecstatic union with God, was the touchstone for medieval mystics. It was what you aimed for, going through intermediate stages of the sort that Evelyn Underhill proposed along the way. Prayer, penance and fasting were regarded by most monastics as essential precursors to gaining 'true' mystical experience, though in fact, as they probably knew themselves, the experiences are often spontaneous. Prayer and the rest seem to play more of a 'practice makes perfect' role than that of generator of experience *de novo.* There's some change of emphasis nowadays concerning how to go about achieving mystical experience due to the influence of Buddhist meditators, who generally aim for loss of all sense of self and being, and for 'pure' awareness with no phenomenal content (other than awareness of being aware).

On the whole, introvertive experience is rated higher than extrovertive by these standards. The *via negativa,* too, if not exactly better, seems to have been regarded as a surer perception of the Divine than *positiva* experience, perhaps because it was less prone to result in the mystic later preaching alleged heresies based on their particular encounter with God. I think this sort of value rating may be theologically dubious because there's a lot going for the idea of a God who is as much immanent in nature as transcendent. The extrovertive experience of a transfigured nature may in fact be just as significant as experiences regarded as 'higher' or 'purer.' But that's by the way. What is generally regarded as the goal of mystical experience is loss of the self in God for Christian mystics or in some formless Ground of Being for Buddhist meditators. How the goal is conceived is down to memes, but the experience itself appears to depend on a supra-personal and even supra-cultural attractor. It's an attractor that has a lot in common with a Jungian archetype; it manifests throughout humanity and takes a wide variety of forms (what Jung called 'archetypal representations'), but retains a distinctive, recognizable identity.

We're back to the old question: is mysticism down to some feature of our biology (presumably the biology of our temporal lobes and related areas), or is it something quasi-mathematical and 'Platonic'? While I suspect that this may prove to be a false dichotomy in the last analysis, it is nevertheless one that is seen as useful by many people. Put the whole thing down to biology and you can look for physical causes of mystical experience following Michael Persinger's lead, or you can concoct 'Just-So' stories about why we should have evolved so as to be capable of it — as some evolutionary psychologists have done. Maybe it's a 'spandrel' that is, a useless by-product of some aspect of brain function that does have evolutionary value; or maybe it's useful in helping us to tolerate the slings and arrows of outrageous fortune, so aiding our survival. You can go on making up such stories till the cows come home. Put it down to something transcendent, on the other hand, and you can go off into realms of theological speculation for just as long.

The idea that we need to hold on to here is that it's an *attractor,* and we've already found reasons for thinking these to be law-like and to possess transcendent aspects (as in the case of the black pumas). What sort of attractor is it? It could be one deriving from our genetics (back to the Persinger-type view again). On the other hand, it's not based in Level 1 because, although memories and perceptions enter into the phenomenology, it's not down to them alone. Level 2 is similarly out because, as we've seen, it's a supra-personal attractor while Level 2 is all to do with conscious personhood. It also seems a bit too universal to be due to an attractor in the social dynamic; moreover it doesn't appear to be wholly attributable to memes despite the contribution they make to it. Are we stuck, therefore, with an entirely 'biological' basis for mystical experience, or could there be a dynamic capable of harbouring attractors yet further up the hierarchy? It's an attractive possibility for it opens up whole vistas that would be forever closed off by a strictly biological view. It's worth remembering that any such level in the hierarchy could be expected, like the others, to feed back and influence those lower down, including the biological one.

Each level in the hierarchy of dynamics emerges from the one beneath. Maybe there's a dynamic of all natural dynamics (which would include our social dynamics) sufficiently coherent to be capable of itself harbouring attractors — including the one apparently responsible for mystical experience. This may seem a far-fetched idea. Seeing that it might not be totally unrealistic will involve taking a long detour in the course of which we shall need to get to grips with very fundamental questions. But first, I want to tell you a fairy story.

A fairy story

Once upon a time, there was a little attractor that lived in a brain — let's call her Smith. The brain was Smith's world and provided her with all she needed — existence, energy and a purpose in being. She had in fact been called into existence as a consequence of certain inputs reaching the brain, and became re-activated whenever similar ones arrived. She was a Level 1 attractor but did not know that, of course. However she did know a certain amount about neighbouring attractors. Sometimes they would help her to grow strong and big; sometimes they would drain her of energy and she would fade away. Occasionally they would help so much that she would grow bigger and bigger and come into contact with all sorts of strangers living in every part of her world. On those occasions, it was as if she had been given a magic broadband connection that allowed her to talk to everyone all at once. She liked that. And she liked helping her neighbours to do their own thing, whenever she got the chance. She was a good little attractor.

Despite being so good, she was nevertheless a teeny bit selfish. She kept hoping, each time she came awake, that it would be her and not one of her neighbours who would get the broadband experience. So she paid special attention to how it arose. It usually seemed to need a whole group of neighbours getting together and trying as hard as they could. Then one of them would suddenly get the reward, maybe followed by others in turn. But Smith couldn't help noticing that some days it all happened easily, with no difficulty at all, and other days they just couldn't seem to manage it however hard they huffed and puffed and pushed and shoved. This puzzled her. 'Surely it ought to work just the same each time' she said to herself. 'We're doing everything that can be done and we always do our best, so why don't we always get the same result?'

She puzzled and puzzled over the problem, but couldn't seem to work it out. Then one day she asked a wise old neighbour, named Jones, if she knew the solution. 'Ah,' said Jones. 'That's the Great Attractor you're talking about. You can't see it and you can't hear it and you certainly can't touch it, but it can make life easy for us or it can make life difficult. You have to be careful of the Great Attractor. Best not to upset it by questioning its ways.' Smith took her advice to heart and was always careful to think respectfully about the Great Attractor, even when it was putting difficulties in her way and preventing her and her friends from reaching the broadband.

Neither Smith nor Jones ever found out that their own efforts were partly responsible for the very existence of the Great Attractor. And the human whose brain they lived in never did realize how much care they were taking to avoid upsetting him. You don't necessarily receive much thanks, if you are an attractor.

Let's find out if a Great Attractor could exist in relation to us Level 2 beings. The social dynamic, of course, acts a bit like one at times. However it's not what we're looking for as it is not mysterious in principle, and it's hard to see how it could engender mystical experience. In materialistic societies like our present one, one would expect the dynamic to do the opposite; namely to inhibit mystical experience. But the experiences occur anyway, and as far as we know are just as frequent nowadays as ever they were. They may appear to be less frequent, judging from literature and the like. But the appearance seems mainly, and possibly entirely, due to our reluctance compared with medieval people to acknowledge or discuss them (though earlier peoples were also often reluctant, if only for fear of being accused of heresy!). Surveys and anecdotal evidence both indicate that mystical experience is alive, well and surprisingly frequent among apparently secular people, despite the lack of any common expectation nowadays that it may occur.

Possibly the biggest hurdle to be crossed before endorsing any such concept has to do with time and causation. Any Great Attractor would have to be regarded as an emergent property of large-scale dynamics everywhere, including our own. The time scale required for emergence would thus be long. From our point of view, therefore, it might have to be regarded as belonging more to our future than to our present or past. So there's no way it could influence our present behaviour or experience is there? Things in the future do not and cannot influence the past, can they? To suggest otherwise is teleology, the idea that the natural world has purposes towards which it aims, which is even more of a scientific heresy than was Lamarckian inheritance in the days before people realized that this does have a role (in epigenetic, behavioural and cultural evolution) after all. Similarly, referring to consciousness in the days when Behaviourism ruled was a quick route to losing all credibility and funding if you were a psychologist. Both Lamarckianism and Consciousness Studies lost their heretical status when people came to adopt broader views of what evolution or psychology are. The broader view needed in relation to thinking about teleology concerns the very nature of time. We'll take a look at that in the next chapter.

13. Time and the Average Person

The idea of time lodged in most people's mind is that articulated by Isaac Newton in the seventeenth century, though it has probably always been around in vaguer forms. It comes in three parts. There's an unknown, unreal future, a present where everything becomes real for a moment, which then turns into a sort of half real, partially knowable past. The progression from future to present to past is a universal process, occurring at the same rate for everyone, everywhere. As the hymn has it, 'Time, like an ever-rolling stream, bears all its sons away.' The beautiful Anglo-Saxon metaphor for our lives, of a bird flitting from the outer darkness into a lighted feasting hall for a moment, then passing on back into the darkness, would equally fit our image of time. It's a notion that serves us well. It's what time is like for us.

There are obvious problems with the picture. For instance what constitutes the 'present' is hard to pin down. Newton himself regarded time, like space, as infinitely divisible. The whole idea of the differential calculus, which he invented (along with Leibniz), depends on there being infinitesimally small increments of time. These figure in equations using the modern notation as 'dt'; Newton himself had an idiosyncratic notation, and the one we use nowadays is due to Leibniz. However the present that we experience is certainly not like that; it has substantial duration. William James, the great American psychologist who was a brother of Henry James the novelist, introduced the notion of the 'specious present' — that is, the period of time that we experience as 'now.' Its duration is very variable, ranging from a minimum of around one tenth of a second when we are extremely alert and aroused, to ten seconds or more if we are relaxing comfortably in an armchair listening to music maybe.

In fact thinking about music helps with trying to clarify what we mean by the present. And the more you try to pin it down, the more wriggly and elusive it appears. If you listen to music in one way, individual notes each seem to occupy their own moment of presence; listen to it another way and entire musical phrases, containing many individual notes, can appear to occupy a single moment. Jason Brown, a New York neurologist, has given us a lovely image of our experience of time being like a fountain, surging up and scattering evanescent drop of consciousness which constantly fall away and disappear. His idea is based on the fact that all 'things' can be regarded as temporal cross-sections of processes; even a rock is a process if viewed on a geological time scale. Consciousness, for Brown, is a wavelike process of actualization of reality.

Anyhow, fuzzy though it may be, our notion of time works well enough in relation to our experience of the world in general. But it is known to be wrong. Like so much of our experience, it's a sort of illusion that corresponds with what the world is really like enough to get us by — but no more than that. It has been known for sure to be wrong ever since Einstein formulated his special theory of relativity in 1905 and his former tutor, Hermann Minkowski, gave it its present mathematical form a few years later. We'll take a look first at what relativity has to tell us about time, and then move on see whether quantum theory has anything to add to the picture.

Space and time

Minkowski gave a famous lecture in the course of which he averred that we should in future no longer talk of space and time as separate entities but of a four-dimensional 'space-time.' His idea stuck. We all pay lip service to it, but what does it actually mean? After all, time is not particularly like space. We can move about in space easily enough, backwards as well as forwards, quickly as well as slowly. Time, on the other hand, is a one way ticket that appears to move us — and usually at a rate that we would choose to alter if we were able.

Moreover Minkowski's mathematics treat time differently from space, for they give it the opposite sign to the three spatial dimensions. If the spatial dimensions are regarded as positive then time is negative — or vice versa (it doesn't matter which way you assign the positivity and negativity). One consequence of this is that, while a straight line is the shortest distance between two points in ordinary 3-D space, it is the *longest* distance between two points in Minkowskian space-time.

This is what underlies the famous space-travel 'paradox.' If you go in your space ship to Sirius, twenty light-years distant, at nearly the speed of light, the return voyage may seem to you to take only a few months. However, when you get back, you will find that all the stay-at-homes have aged by forty years. Despite travelling further in ordinary space, you in fact took a relatively short path through space-time compared to the stay-at-homes. So, although you didn't notice any subjective feeling that time had slowed down for you while you were on your journey, in fact it did move more slowly for you than it would have done if you'd stayed home. Those left behind travelled a longer space-time route while you were away, so they aged more.

And the mathematics are right. Time does appear to us stay-at-homes to move a lot slower for very speedy objects, such as subatomic particles in particle accelerators, than for us. The longest space-time distances of all — that is, the straightest lines — are to be found in areas free of gravitational fields (so *general,* not special, relativity tells us). Lo and behold, atomic clocks in orbit about the earth tell time differently from identical ones down here on the surface with us. They are a tiny bit slower because they are moving relative to us and a tiny bit speedier because they are in a lower gravitational field. Both effects have to be taken into account in relation to Global Positioning (GPS) satellites for instance, because they don't usually cancel out.

What if you were to travel to Sirius at the speed of light? You could not in fact do so because other bits of special relativity show that your mass would increase towards infinity as you approached the speed of light. And of course there could never be enough energy to continue accelerating a mass that was behaving that way. However, photons travel at the speed of light for they *are* light. If you beamed a photon at Sirius and it was reflected back to you on arrival there, presumably no time at all would have passed on the round-trip journey from its point of view — but still forty years from yours. And if it went a bit further before being reflected, say halfway to the limits of the visible universe, then everything that has ever happened in the history of our bit of the universe would have occurred in what would seem to the photon no time at all.

Of course it's misleading to regard photons as having a point of view or a possibility of experiencing the passing of time, but there is a serious point here. The apparent passage of time in bits of the universe that you observe depends on several factors and, in principle, there would seem to be a point of view from which an entire universe could appear to evolve and fade in what would seem like an instant to you.

Observers don't necessarily agree on the rate at which time passes and they don't always agree on the ordering of events either — at least they don't if they rely on the evidence of their own senses. Let's say that tensions between Daleks and Sirians, who occupy neighbouring star systems exactly two light years apart, become so great that they launch their (sub light-speed) space fleets in what they intend as pre-emptive strikes. When each discovers what the other has done, they find there has been an amazing and disastrous coincidence. Both first see their opponent's launch two years to the day after their own. They must have taken action more or less simultaneously, they conclude, and will probably be annihilated at the same time too, given that their ships use equivalent technologies.

Captain Kirk however, coming out of warp drive but still moving at about 98% light speed, happens to be on a trajectory from which he perceives the Daleks to have launched their fleet five months before the Sirians did so.[1] 'Typical Daleks!' he thinks to himself, 'They are the aggressors as usual and the poor Sirians are simply defending themselves. Starfleet must be told to aid the Sirians!' Meanwhile a Klingon ship, moving on a very different trajectory but also at a significant fraction of the speed of light, draws the opposite conclusion. Its captain clearly observes that the Dalek launch was no more than a response to that of the Sirians, which had occurred three months previously from his point of view. Earth/Klingon tensions mount to danger levels as they offer support to their respective 'victims,' but luckily Mr Spock then does the appropriate calculations and proves to both captains that they were wrong.

Considerations like these have led many philosophers to adhere to what they term the 'block universe' view of time. Not all philosophers agree, of course — it would be a miracle if they did! But the block universe view clearly has a lot going for it. What it says in essence is that all times, future, present and past, are equally real (or unreal), and our perception of movement through time is purely a function of our particular perspective on it. It's a view that seems a natural consequence of special relativity in particular, and there's plenty of experimental evidence that the theory is right. However, just to introduce a note of caution, relativity theory (both special and general) can lead to nasty problems and apparent paradoxes. For instance, there are all sorts of difficulties to do with black holes. One of these is connected with their so-called 'event horizons.' If an observer were to watch something falling into an event horizon, it would appear to her to become sort of 'smeared out' towards infinity the closer it got to the horizon, but would never actually cross it. If she were to fall through the horizon herself, however, she would notice nothing unusual (provided the hole was large enough for tidal effects to be negligible). The underlying physics would seem to be very different according to whether you are falling into a hole or simply watching something else fall — and physics shouldn't be like that! Any resolution of event horizon paradoxes will probably require new physics. And who knows what sort of picture of the universe and of time will be entailed by the new ideas, when they are available?

Some have objected to the 'block universe' picture on the mistaken (but perhaps more understandable) grounds that it rules out 'free will.' I'd like to deal with this objection before moving on. It's a bit of a side issue to the main theme of this chapter, but is often felt to be a stumbling block by people pondering over alternative concepts of time. Their argument is similar to that offered by Calvinists for predestination. If God knows everything, then he must be able to see that you are going to choose to become a sinner and end up in Hell. And if he can see that happening, you must inevitably make the choices that will lead you to Hell. So the 'choices' can't have been free. You were always destined to make them, no doubt due to God's inscrutable providence, and were bound to end up frying. Similarly, if all time is equally 'real' then everything you will do is already 'there' in some sense, so you haven't in fact got any choice about what you'll do ... right? No, sorry, it's wrong!

Actually God watching you do something is not the same as God making you do it. C.S. Lewis of 'Narnia' fame pointed this out long ago. You could have chosen differently, it might be supposed, in which case God would have seen a different scenario and Calvinists, well known for being privy to God's insights, would have claimed instead that you were always destined for the pearly gates. Equally, if you choose now to go left instead of right, the block universe will contain you heading off to the left. On the other hand, if you choose right, there you will be eternally going right. The block universe picture in fact has no implications for 'free will,' neither ruling it out nor supporting its existence.

Getting back to relativity, one big thing that it's no help with whatsoever is to do with why our experience of time should have a preferred direction. The maths of relativity theory work just as well backwards as forwards. Therefore it can't tell us, people sometimes say, why the flow of time is from future to past instead of vice versa. But putting the problem like that is based on thinking of the flow of time as being like that of a river. You could tell if a river went into reverse because you could watch the trees growing along its banks or whatever. They would look as if they were moving in a different direction. But there's nothing outside time to give any directional coordinates for its flow. Therefore, we should have no way of telling if it *did* reverse from the perspective of a God who was outside time — everything would simply seem to continue flowing as before. The right question is to do with what differentiates past from future in our perceptions. The short and personal answer is 'memory'; we remember the past, but we don't remember the future — the present is the point at which the memory process initiates. The more general answer that applies to the whole universe is 'entropy.'

'Entropy' is a concept from thermodynamics and is a measure of the amount of disorder in a system. In closed systems it is bound to stay the same or increase. In open systems, like our bodies and brains, it can and does decrease though at the expense of exporting extra entropy to some other part of the environment. The ecosystem of our world, with all its amazing complexity and order, depends ultimately on importing relatively low entropy short-wave photons from the sun and exporting high entropy infra-red radiation into space. The universe in its entirety is actually the only perfectly closed system that we know about, for its separate parts, however apparently isolated, always have *some* connection with other parts. This means that, at least in the universe as a whole, entropy is bound to increase with time (staying the same is not an option for such a dynamic system as the universe). The past is the direction in which entropy is lower and the future that in which it is higher.

While the entropy explanation is the one normally given to account for the direction time's arrow, there's another hovering in the wings which has never, so far as I know, been adequately formulated. It's to do with asymmetrical processes associated with the so-called 'weak nuclear force.' Physical processes generally are indifferent to direction in space. As mentioned earlier (Chapter 2) the law of conservation of momentum is down to this. However, these 'weak' processes do have a small preference for 'left' over 'right.' It's been hypothesized that the preponderance of matter over antimatter in the universe may be due to this preference. The asymmetry also involves time, so it's just possible that the direction of time's arrow has something to do with the processes that made us from matter instead of antimatter.[2] However this is speculative and uncertain, while the 'entropy' explanation is as solid as anything can be in this confusing field.

Entropy notwithstanding, Richard Feynman, that giant of twentieth century physics, pointed out that positrons (the antimatter equivalent of electrons) can be regarded as electrons going backwards in time relative to us. Indeed he and his even greater colleague and former mentor, John Wheeler, played for a while with the notion that many quantum phenomena are due to interaction of waves travelling forwards in time in the normal way with others travelling backwards to the point that we would think of as their origin. It never really caught on possibly because it seemed just a little too baroque (though it pales into insignificance in this respect compared to string theory with its ten or eleven dimensions and near infinity of different variations). Also it simply didn't seem to fit with the sorts of causation that physicists like to consider. After all, you can't break a window today by throwing a stone at it tomorrow. On the other hand, the idea does have a lot more going for it than you might suppose. Australian philosopher Huw Price concluded towards the end of his brilliant book (see Further Reading) that quantum theory is just the sort of crazy picture you could expect if influences *can* travel back in time.

Actually quantum theory is quite 'schizophrenic' when it comes to time. The famous wave equation at the basis of traditional quantum mechanics contains a Leibnizian 'dt' as one of its terms. This means that it is using a *Newtonian* concept of time, which is not only wrong from the point of view of relativity theory but also from that of quantum theory itself! For time is *not* infinitely divisible in quantum theory. There is a minimum duration known as the Planck (after physicist Max Planck) time. This is so incredibly brief that it makes no difference for all practical purposes; but there is nevertheless something a little crazy in principle about using the 'wrong' notion in the maths of your theory. Modern quantum field theory has in fact incorporated special relativity successfully, though general relativity (the extension of the special theory that accounts for gravity) has so far proved wholly indigestible. But the 'dt's still lurk within much of the maths.

All the same quantum theory does have interesting things to say about time, in two main respects. First, it reinforces the 'entropy' picture of time's arrow, for whenever a quantum wave function collapses there is an associated entropy increase,[3] and such collapses are irreversible in principle as well as in practice. Secondly there is 'quantum entanglement,' the implications of which are astonishing. Let me explain.

Alice and Bob are the good guys. They need to exchange information that can't be read by evil Eve. Actually this is impossible. Eve can always eavesdrop somehow, at least in principle if not in practice. But, if Alice and Bob are quantum cryptographers, they can ensure that they will always know whether Eve has been prying into their messages. How do they manage this?

One of them, Alice as the story is usually told, makes pairs of entangled particles. There are lots of ways to do this; it has become routine in laboratories investigating entanglement. She keeps one particle from each pair and sends the other to Bob. Then she makes some measurement on each of her particles, chosen at random from a range of possibilities, and phones

Bob with two pieces of information; first to let him know what particular measurement she made and second what the outcome was. Bob measures his particle in the same way, then compares his result with Alice's. Provided no-one has eavesdropped on Bob's particle, his measurement will always be perfectly anti-correlated with Alice's. If Eve did sneak a look at Bob's particle on its way to him, the measurements won't always agree. So Bob's particle can both carry a message from Alice and can help tell him and Alice whether anyone intercepted the message.

Entanglement can be useful, to be sure, but is also very weird. One of the weirdest aspects is the way it seems to ignore time. The Alice and Bob scenario would still work in principle if Alice was an earthling and Bob lived in Andromeda, two million light years from earth. Suppose Bob were to receive a particle that had originated from earth, and Alice happened to come into possession of its entangled partner, then Alice's measurement would still be perfectly related to Bob's provided nothing had interfered with either particle in the meantime. Probably the easiest way of envisaging what might happen is to think of possible timings:

2,000,000 BC — Creation of entangled particles. One gets trapped on Earth. The other heads off for Andromeda

AD 1 — Bob (a long-lived Andromedan) receives and stores his particle.

AD 2010 — Alice measures her particle and transmits the outcome to Bob.

Entirely by coincidence Bob measures his particle in exactly the same way and at a time that could be regarded as one minute later, if instantaneous message transmission between Earth and Andromeda were possible.

AD 2,002,010 — Bob receives Alice's message and then gains information about whether his particle was probably untouched before he measured it.

There's no problem with the classical information here, which flows in the usual orderly way at light speed or less. But something information-like can be pictured as connecting Alice's measurement with Bob's, even though Bob is two million light years away. This connection fixes the result Bob gets (provided nothing has interfered with either particle in the meantime) when he measures the particle which he stored long before Alice was born.

In practice the above scenario is unrealistic if only because the particle heading for Andromeda would be sure to interact with something on the way, such as the cosmic microwave background, and would thus be ‘interfered with’ on arrival. But the principle is sound enough and has been experimentally proven over distances of several kilometers. Entangled particles react instantaneously to one another. In fact they have to be regarded as constituting, in a sense, single objects however far apart they appear to be situated in space-time. What does it all mean?

The first thing that commentators usually hasten to point out is that quantum entanglement can’t be used to transmit messages instantaneously. These have to go at the normal light speed or less. Star Trek-type chat with a base light years away is not possible. Some then go on to add that ‘information’ can’t be transmitted instantaneously by these means. However, they are wrong if they do so. What they should say is that ‘meaning’ can’t — that has to await the light speed or less message from Alice. ‘Information,’ however, is usually taken to mean ‘Shannon information.’ And Claude Shannon defined a ‘bit’ of information as the answer to a single yes or no question, regardless of what the yes or no meant. Clearly Bob has this answer as soon as he makes his measurement, so he has the information. It’s just that he has no idea what the information means until he hears from Alice over normal channels. We’ve been using the notion of Bateson information (a difference that makes a difference) in this book. But the same applies in this case. Bob’s measurement makes a difference — to his measurement apparatus and to his knowledge of the particle — though again the information can convey no meaning till Alice’s message arrives.

The next thing commentators often get to is the fact that entanglement is an almost universal phenomenon. Any two particles that have ever interacted must be regarded as having become entangled. Eve’s interference doesn’t totally destroy the entanglement that Alice set up. Eve’s action merely swamps Alice’s entanglement in a sea of new entanglements, so that Bob can no longer access the information that his particle originally contained. He cannot isolate Alice’s entanglement from the host of others that Eve’s interference brought about — a flood that encompasses the entire history of the particles that make up whatever instrument Eve used to measure Bob’s particle, right back to when they were formed in some star maybe. The thought that the whole

Universe is a vast sea of entanglement is so mind-blowing that people often give up saying more about it. Some brave souls, however, put that aside and go on to point out that the phenomenon must probably be taken to imply that neither space nor time are fundamental properties of the universe. They must derive from something else whose nature we can only guess at.

I'd like to focus on a rather different implication here. It is that, because Bateson information is conveyed between entangled particles, there must be a universal dynamic which is not subject to relativistic limitations. Differences that make differences are at the basis of any dynamic and they can be transmitted instantaneously between entangled particles, it seems, albeit only when divorced from meaning. And entanglement relations exist everywhere. Therefore a strange, ghostly dynamic must exist everywhere. Could it harbour attractors? There's no way of knowing for sure, but it's highly probable the answer is 'yes.' They are properties of dynamic systems in general and there is no reason to suppose that an entanglement-based one should be any different.

Someone might object that it makes no sense to talk about the dynamics of an apparently timeless system, like the one envisaged here. Surely you can't have movement and change without any time to move or change in. And they would be quite right to have doubts. The overall picture does indeed have to be viewed as a structure rather than a dynamic. It's another landscape — like an attractor one. However, from our human point of view, it would look as though it had a dynamic if we were somehow able to observe it directly (which we can't). It would appear to evolve as we inched forward in time over its 'surface.' And our own actions and choices would contribute their miniscule quota to what we would perceive as its evolving structure.

The structure itself, however, is independent of our time-bound perceptions. This means that aspects of the structure which would appear to us to derive from occurrences in our future could influence our present or indeed our past, if they are capable of affecting us at all. I'm imagining here that a Godlike point of view exists, able to see both the overall structure and our position on it as we make our time-like progress over its surface. From a time-bound point of view on the surface of the structure, we could be imagined to participate in events whose consequences later (from our point of view) affect the dynamics of entanglement and thus the shape of the landscape. However, from a Godlike point of view, the new shape is already there and possibly already doing its thing in what we regard as our present. That's where teleology may come in.

The next question, therefore, is whether this dynamic could have any effect on us. We know that physical causation is both strictly relativistic and one-way only in relation to the arrow of time. You might think gravity could spread instantaneously because it is a distortion of the shape of space — and surely distortions like that can 'flip' all at once. If you did think that, you'd be wrong. Even gravity is confined to light speed. If the black hole at the centre of our galaxy were suddenly transported now to the other side of the universe, we should not be in any way affected for another ~30,000 years.

All physical causation is subject to the time of relativity theory, so attractors in an entanglement dynamic could never have any physical effect on us, could they? Hang on! Attractors aren't like physical causes anyway; they are law-like. They're outside the remit of relativity, and are thus not barred from affecting us even if they exist only in the ghost-world of entanglement, not in the 'real' world of neurons and meaning. Equally the actions following from our choices will affect them, for actions inevitably set up new patterns of entanglement, each of which will make its own tiny contribution to the overall dynamic.

Let's pause at this point and take stock of where we've got to.

We've seen that relativity theory can be taken to mean that, in relation to time, we're like a group of nearly blind ants all trudging together at a constant pace along a road. We can't see what lies ahead, but we can see a little of our immediate surroundings and remember a certain amount about where we've been. Every now and again we meet beetles and centipedes, who sometimes move at different speeds from us but always in the same direction. There are rumours of creatures able to go the other way, but they remain no more than rumours. The road itself exists, apparently, whether it is ahead of us, under our feet or behind. On the other hand, we do sometimes arrive at forks in the road, when our choice of route determines which parts of the landscape we shall encounter in our personal futures.

Quantum theory suggests that, while this picture is true enough from our perspective, the road of time is not in fact as solid as it may seem. Indeed it may be more like a holographic projection than something material. And a communication system exists, it appears, which can instantaneously reach from one section of road to another. A railway metaphor would serve better here, for what the communication system does is to reach ahead down the route and adjust an interim destination, like switching the points — so that when us ants finally get further down the line we'll find ourselves going left instead of right, for instance, or coming up with the answer 'yes' instead of 'no.'

The ants may be able to influence their own future route in two ways. There is direct causation. It is as if they carry lengths of railway track that they lay as they proceed, and are sometimes able to select in which direction to lay them from a choice of alternatives. In addition their track laying may, it seems, influence what is sent on down the line; an influence which can be thought of as continually re-modeling their potential futures in advance of them arriving there.

What about the possibility of teleology, raised at the end of the last chapter? As we've seen in this chapter, there's nothing in our modern understanding of time to rule out some form of it. The 'block universe' view suggests that the future is in a sense just as 'real' as the present or past. And what's real may well have effects of some sort. If the future does have effects, they will be mediated through consequences of quantum entanglements. These, once formed, appear to be outside time altogether, in the sense that they are not affected by the constraints of relativity. So it's very likely that they evade time's arrow completely. From our viewpoint, therefore, influences may appear to go backwards as well as forwards in time. As Richard Feynman once said, *apropos* all the phenomena of the quantum world: 'Anything that is not forbidden is compulsory.' If entanglement influences are not forbidden from appearing to travel backwards in 'our' time, they are likely compelled to do so.

We can detect some entanglements when they leapfrog into the future, as in the Alice and Bob scenario. But we have no possibility of directly seeing them if they sometimes jump from the future into our present, or if they jump from our present to our past. But, according to Feynman's dictum at least, they are likely to do so. Presumably any particular entanglement could not jump back to a time before it formed, but its effect on the overall entanglement dynamic might not be so constrained. In any case, the ones we're interested in from the attractor dynamics point of view almost all belong to the 'sea' of entanglement that exists out there, not to very special, isolated set-ups in laboratories. Many of the 'sea' entanglements date from quite soon after the 'Big Bang' in which the universe was formed, so any constraint on how far back their atemporality extends is unlikely to be significant from our point of view.

If a Great Attractor exists (or indeed any lesser ones situated at the same level of the dynamic hierarchy), it has to be in the entanglement realm. So teleological effects will proceed from the same area. They won't manifest in any form of physical causation. All the physical causes associated with any manifestations will be of the usual sort and will derive from the physical bases of dynamics further down the hierarchy — interactions between people or between neurons or maybe even the chemistry of genes. Manifestations will be in law-like influences that appear to constrain physical causes; maybe also in emergent statistical effects and even in things like spooky coincidences of the types that fascinated Jung — he coined the term 'synchronicity' to describe them. We'll be thinking a bit more about where one might look for evidence of their occurrence in the final chapter.

Perhaps something should be said about the theology of this concept, for a Great Attractor could certainly be envisaged as God. Theology's not a field I know anything much about, so I shall say very little. The Attractor is an emergent property of the universe, and I believe that the concept of emergent deity has some currency in theological circles. It would seem to have both advantages and disadvantages.

The advantages are that it offers a route to imagining how God could be integrated with nature, which must surely be of interest in our current, non-dualistic, intellectual climate. Much more important, it would appear to solve the theological aspect of the problem of suffering. An all-powerful God who creates, or at best permits, torture and the death of children from painful disease seems to me one that nobody in their senses would wish to believe in, and certainly not to worship. All the reasons most of us were fed as children, about how He might create or permit suffering in the interests of our moral development or our 'free will' or whatever, are surely nothing more than rationalizations — and poor ones at that. The idea of 'Karma' as an explanation for suffering, payback for misdeeds in past lives, seems little better. One might accept that *some* suffering is good for us, for clearly it occasionally is, and that free will is bound to have its downside. But so much pain; and much of it so very cruel? No way!

An emergent God, however, has no choice in the matter. He's a product, not a permitter, conniver or originator of suffering, just as much as he's a product of all the other aspects of the universe that enter into his make-up. He is, in a sense, a suffering God, though suffering for a God emergent from the entanglement realm might not have meaning. As we've seen, information appears to be divorced from meaning that realm. It's we, in our sub-lunary, relativistic world, who endow the universe with meaning according to this picture. I used to wonder as a child, listening to the lessons read out in church, why Jesus, if he believed he was the Messiah that he claimed to be, should have so often been reported to call himself the *Son* of Man. Maybe it was simply mistranslation, or shorthand for something like 'Son of God sent to Man' On the other hand, perhaps it is an accurate report. If so, there's an explanation of a sort implicit in the line of thought we've been following here. If his belief in himself was valid and he was indeed an aspect of an emergent God, then he was a *product* of man and womankind along with the rest of the relativistic universe.

On the other hand, an emergent God is neither all-powerful nor a good candidate for creator of the universe. Maybe He's not such an attractive proposition theologically after all. Then again, possibly the universe is like the *Ouroboros,* the cosmic snake that eats its own tail. In that case, the God that the universe creates might create the universe. I've no good ideas to offer about this, so best to leave it there.

14. Uses and Tests

The central idea that I've tried sketch out for you in this book is one that links together every aspect of our minds from their genetic underpinnings through neurology to sociology and beyond. As we've seen, it offers a conceptual framework for thinking about most aspects of our nature, including perspectives on truth, creativity, beauty and more. But what use is it? How can it be tested? These two questions are interrelated. If the model turns out to be useful, the implication has to be that it possesses some sort of validity. Moreover, if it passes tests of validity, uses are likely to be found for it. First, here's a summary of the whole thing:

Atemporal entanglement dynamics
(based on particle entanglements)

Social dynamics
(based on memes)

Level 2 dynamics in brains
(mostly conscious:
based on interactions between Level 1 attractors)

Level 1 dynamics in brains
(mostly unconscious:
based on interactions between neurons in neural nets)

Brain genetic dynamics
(based on development processes guided by genes)

- There are linkages running both up and down the hierarchy.

- Links proceed up the hierarchy step by step, except for the top stage (entanglement dynamics) which may be directly affected by all lower stages.

- All ascending links are mediated by ordinary, straightforward physical causation.

- Descending links do not necessarily go step by step; for instance, 'social dynamics' may directly affect the bottom level ('genetic dynamics') as well as the two levels immediately below it.

- Descending links often appear to be mediated by law-like influences, which act in conjunction with ordinary physical causation as far as the four lower levels of the hierarchy are concerned.

Evidently the most 'cloud-castle-like' component of my model is the top of the hierarchy, 'entanglement dynamics.' I included it because I believe that any picture of mind which omits it is likely to be seriously incomplete. However it does complicate the task of finding evidence and uses for the model, since both are hard to come by in its case — though not entirely inaccessible, as we have already seen ('black pumas' and the like). Because it is rather different from the rest, I'm going to postpone tackling it until the final chapter and will look at the other components in this one.

There is probably no way of validating the model as a whole, even when the top level is excluded. It is like evolution in that respect. You can't prove that Darwinian theory as such is correct. All you can do is show that lots of facts fit in with it, and that sensible predictions about the detail of biology can be derived from it. Similarly, with the attractor dynamics picture, we need to look at some of the detail plus explanations and any predictions that can be derived form it. If these stand up to scrutiny, this gives some confidence that the entire concept may be worthwhile.

Perhaps the first thing to note is that the overall behaviour of each level is similar to that of the others. Starting from the bottom, genetic dynamic level, it is now fairly generally accepted that Stephen Jay Gould's picture of 'punctuated evolution' was often right. There are long periods of relative stasis, followed by bursts of change. Some changes seem to be precipitated by environmental catastrophes — asteroid impacts or whatever. Other changes are thought to arise from the dynamics of the system itself. Maybe the original explosion of multicellular life in the Cambrian period was like that. On the other hand the two types of cause are hard to disentangle, for changes in the system produce changes in the environment harbouring the system; as when the early photo-synthesizers generated oxygen that was poisonous to all extant life forms, resulting in a cascade of adaptations ending up with trees and us. As a consequence of this difficulty there is argument over a wide range of 'chicken and egg' questions, such as whether the dinosaurs were already in decline and the asteroid impact often blamed was no more than a final straw causing their extinction.

On much shorter time scales, similar patterns of alternating stasis and change can be seen in brain (at both levels) and social dynamics. Everything goes along as normal for a while, then something happens and big changes ensue, followed by all sorts of secondary alterations. The 'something that happens' may originate from the environment. You may be driving peacefully along, for instance, day dreaming about your next holiday, when suddenly a lorry jack-knifes in front of you. That has all sorts of immediate effects on your brain dynamics at every level, and you are very lucky if it doesn't seriously damage the rest of your dynamics too. Even if it doesn't, your brain dynamics will remain disturbed for quite a while. On the other hand, dynamic changes can occur for no obvious reason. Ambiguous drawings — the vase/faces, duck/rabbit/, young girl/old crone, and so on — are good examples. If you look at them long enough, what you see flips from one of the alternatives to the other of its own accord, and goes on flipping indefinitely. That's mainly down to a Level 1 happening, presumably, but similar changes happen at Level 2. You can switch from thinking about a problem at work to thinking about the new car you would like to buy, for no reason outside your own brain dynamic.

Social dynamics show the same pattern. Everything goes along smoothly for a while, then there is an upheaval and everything changes. The upheaval may be due to a natural catastrophe, like the explosion of the Santorini volcano that allegedly did for the Minoans, or it may be due to social events outside your control, for instance, someone invading your country. On the other hand, change may be due to internal dynamics as exemplified by the French revolution and, arguably, the collapse of the Roman Empire. Large-scale alterations can be surprisingly fast. Living in Western Europe in AD 375, for instance, life must have seemed pretty much business as usual. 'Sure the empire has problems, but nothing it hasn't coped with before,' people no doubt said to one another. Thirty years later they would not have said that. Shreds and tatters of the old order still existed where some particularly powerful senatorial family or bishop held sway, but the overall structure was obviously dying. The consequences of its death reverberated down the centuries. We are still living with them. And the 'chicken and egg' questions are there in full force. Edward Gibbon in the eighteenth century famously attributed the decline of the Roman Empire to the triumph of barbarism and religion, regarded as external forces. More modern historians have tended to emphasize happenings in internal dynamics — the empire's fiscal, educational and social policies.

This commonality of pattern provides an important motivation for proposing an attractor dynamics picture. For it is a pattern consistent with the behaviour of attractors. The 'strange' ones harboured by chaotic systems appear to figure prominently in brain dynamics. 'Periodic' attractors, too, may sometimes show themselves — the 'flips' in ambiguous drawing perception can possibly be put down to these. 'Point-like' attractors occur at Level 1 in relation to particular percepts and habits, but are possibly less frequent in the very open and complex Level 2 brain systems. Point attractors favour closed and relatively simple systems. The genetic system harbours lots of them, for instance, the 'wing' example given in Chapter 1, perhaps because it can be regarded as relatively 'closed' in the sense that genes themselves are stable and mostly immune to outside influences. Social dynamics, too, has plenty of point attractors — living in houses is a possible example — and it too contains stable, gene-like entities in the form of memes embodied in artefacts.

We've already seen that an attractor dynamics picture is acceptable at the genetic level and is fairly 'mainstream' at the neural net level, corresponding to our Level 1. I think it fair to say that few would have strong objections to the picture in relation to Level 2 or to social dynamics, though they might well have reservations about its usefulness, or about how widely applicable it is. But how could links between dynamic levels work? The model is in trouble if there are difficulties here.

Links going up the hierarchy can all be thought of in terms of ordinary physical causation, and they ascend step by step. In relation to brain development, genes specify the production of signalling molecules whose concentration gradients guide nerve cell processes (axons) to grow towards particular targets — that sort of thing. Once the neural nets are formed, the electro-chemical interactions self-organize in a manner modulated by inputs from the environment, resulting in attractor/memories that underpin perception and motor behaviour. The attractors at this level then interact electrochemically, so Henningsen's hypothesis goes (Chapter 3), to give rise to the 'Level 2' attractors that are mainly associated with conscious perception and behaviour. These attractors in a sense *are* the ideas and actions which get translated into the social dynamic via a wide range of complex causal chains — speech, writing, weaving cloth, carving stone, and so on — which all have purely physical bases.

There are no 'in principle' problems with any of this. Nor can details be used to test the attractor dynamic picture, since it cannot be expected to influence them. It affects what happens at each given level in the hierarchy, but doesn't affect upward transmission of the outcomes of what goes on within a particular level. The situation is rather different, however, in relation to downward transmission — at least from the point of view of the entities which exist lower down.

In one sense, there is nothing more to downward transmission than chains of physical causation, as in the upward case. The Roman Empire falls because of its social dynamic, or for some other reason; you, an early fifth century person, suddenly find your environment greatly changed in all sorts of ways. Inputs from your eyes, ears and probably stomach too (the grain stores have all been plundered), tell you about this, which affects your brain dynamics especially at Level 2. Level 2 gets you out searching for edible plants, for instance, which in turn helps to form or activate all sorts of previously unfamiliar Level 1 events. It does so by biasing the electrochemical environment of relevant neutral nets. Then starvation sets in maybe, but you are one of the lucky ones possessing genes better able to cope with that, so your genes have a head start in the evolutionary lottery and the population genetic dynamics is thereby affected.

What would all this look like from a gene's point of view? Let's take a specific example, much quoted nowadays. It's to do with the gene for lactose (the main sugar found in milk) tolerance, which is present and active in all healthy babies as it helps them to digest milk from their mothers. In most populations, however, it gets switched off in childhood soon after weaning. But around 90% of people of European descent keep it active. The generally accepted theory to explain why is that Europeans are like this because they domesticated cattle and took to drinking milk from that source a few thousand years ago. Other peoples who take milk after babyhood, the Masai of east Africa for instance, have the same genetic abnormality. If you don't have it and you swallow milk or milk products, you are liable to get all sorts of stomach upsets. And if milk from cows is an important source of nutrition in your society, you will be poorly placed in the survival stakes if you cannot drink it.

What is down to a memeplex therefore — that is, one about keeping cows and drinking their milk — has affected the genetic behaviour of most Europeans via enormously complex chains and loops of causation. But sentient genes would be unable to see any of this. An Archimedes among them might long ponder the problem before coming up with the solution: 'Eureka!' he would eventually shout, 'I've discovered the law of the survival of the active lactase.'

The lesson to be drawn from this is that many of the downward hierarchical influences will look lawful. And they will *be* lawful in the sense that outcomes will be determined by what goes on further up the hierarchy, not by 'visible' physical causation at the same level or lower. Note that there is no spooky atemporality here. The lawful influences are in fact ultimately down to physical causation, but only in a manner analogous to how the liquidity of water is down to the quantum mechanically determined behaviour of its constituent atoms. Just as it is impossible to predict how water will behave from looking at the quantum mechanics of individual atoms, so it is impossible to predict from the local webs of physical causation alone how anything will behave that is under the influence of factors further up the hierarchy.

The first 'prediction' from the model, therefore, is that social dynamics will influence us Level 2 conscious beings in ways that look law-like to us. And indeed they do, as shown by some of the examples give earlier. Waves of enthusiasm for Mesmerism, Alien Abduction or Neurasthenia have emerged from the social dynamic and swept people up into them willy-nilly. We may put this down to fashion or 'mass hysteria' or whatever, but it is nevertheless lawful in that it affects what people can or will do, independently of their inherent brain dynamics. However many EEGs, fMRIs or other physical factors were studied, it would never be possible to predict from them that a craze for hoola-hoops, say, will recur in five years' time. That's entirely down to something in the social dynamic, which every now and again decrees to teenagers: 'X % of you shall spend inordinate amounts of time exercising thus.'

Maybe this is so obvious that it does not make a particularly good 'prediction' — 'explanation' would be a more accurate term for it. I had better try to extract something a bit less obvious from the model. Two lines of enquiry come to mind. The first has to do with creativity and is also mostly about explanation rather than prediction, though it does have predictive implications; the second concerns the behaviour of attractor landscapes in general and suggests some reasonably well defined predictions. We'll take them in that order.

Creativity in the sense of innovation, so it was argued earlier (mainly in Chapter 11), is a function of the formation of new features in old attractor landscapes. These will be felt to possess aesthetic value if they 'oil' the Level 2 dynamics, we concluded. It follows that new features are more likely to form if the old landscape has been 'shaken up' somehow. And it may not matter all that much how it is shaken. What will matter most is the intensity, and maybe the duration, of shaking. Of course it cannot be too intense or too prolonged because stable new landscape features might be unable to form under those circumstances. It's going to be a matter of not too little, not too much, but just right. And what is just right for any individual innovator will depend on their personality characteristics. Some people find that minor upsets shake them to the core, while others appear unperturbed when the world is falling about their ears. Is there evidence that innovation is in fact aided in this way? One might point to a range of anecdotes; for instance, Newton getting his ideas about gravity after having had to return home to the country for fear of catching plague in the city — he did not find it easy to adjust to his family. But something a bit more systematic would be helpful.

A good example is provided by the huge burst of innovation in psychopharmacology in the 1950s, which has been magnificently well charted by psychiatrist David Healy in his three-volume blockbuster *The Psychopharmacologists*. Lots of individual innovators were involved, so I won't try to detail their histories here. But various themes stand out. The largest group were Continental Europeans, whose countries were in the process of recovering from the Nazi invasions and World War 2. They were *not* people whose lives and research had been progressing along well-ordered lines, cushioned by regular employment in long established hospitals or laboratories. Another group consisted of immigrants, mostly to the United States. The creativity of similar immigrants is of course legendary, especially in fields like physics, and is usually put down wholly to their native genius being able to express itself and flourish in a freer, more supportive environment. But maybe the actual shock of immigration and adjusting to a new country also makes a contribution. A third group consisted of people who had remained in one country but had switched disciplines. And there were a few who had pursued 'normal' (by today's standards) careers, but they were very few and arguably not the major innovators.

As usual in this field, the 'chicken and egg' questions abound. Does immigration or switching disciplines promote creativity or is it that creative people are more likely to follow these paths? It's a bit of both, I guess. The example of apparently ordinary people who appear to have been spurred into innovation by social catastrophes outside their control, suggests that upset as such, regardless of personal qualities, is helpful to innovation. On the other hand some immigrants (especially refugees from the Third Reich), along with the polymaths who switched fields, were clearly exceptional people anyway.

Social upsets are not the only circumstances that might be expected to stir up old landscapes. Mood swings and intoxications might also be expected to do so. Both are known to occur in some creative people, especially writers, more frequently than in the general population. Alcoholism and/or recurrent depression or bipolar swings are occupational hazards of being a successful novelist, it often appears. Many of the romantic poets, Coleridge especially, were famous for their liking for opium and, to a lesser extent, cocaine. Creative artists other than writers also have more than their fair share of moody drinkers and drug takers, though to a lesser extent. Creative scientists, on the other hand, are apparently no more at risk in these respects than people in general.

There's been endless speculation as to why creative genius should sometimes go along with madness, which has become more focused recently as more has become known about the particular relationships that exist. In general terms there may be some relationship between genius and having a relative with schizophrenia, but it's very hard to be both a genius and a schizophrenic yourself — although a few mathematicians and artists have managed it. The so-called 'affective disorders' on the other hand, depression and mania/hypomania, do seem to afflict geniuses more often than the rest of us.

According to the attractor dynamics model, affective disorders should be roughly equivalent to major social upheaval in promoting creativity. It therefore predicts that geniuses whose creativity manifested in the context of invasions, emigration or the like, should on average be *less* prone to affective disorder than other geniuses because (a) they don't need the extra boost to creativity provided by affective illness and (b) a combination of both social stress and affective illness would be likely to prove too much of a 'good thing' and inhibit productivity. This is actually the reverse of what might be expected, since stresses like immigration sometimes cause depression and one might anticipate that immigrant geniuses would be as vulnerable as anyone else. Somewhat counter-intuitive predictions like this are always a plus for any model. Unfortunately I don't know of any sufficiently rigorous, relevant studies, so whether the prediction is valid is an open question. People with schizophrenia, by the way, should be less likely to express genius (as is known to be the case) because the illness can be envisaged to cause fragmentation of attractor landscapes, thus tending to prevent the formation of new connecting features.

But why should successful novelists in particular be so prone to depression/alcoholism? A tentative explanation can be derived from the model. Speech and writing evolved recently compared to most of our faculties. Therefore there has been less time for the biological bases of these verbal landscapes to become integrated with the rest. Anecdotes in Chapter 11 described the difficulty some creative thinkers have experienced in translating their thoughts into language, strongly suggesting an integration problem. Indeed we all know from our own experience how hard it can be to find the right words to express a thought. But if your creativity *depends* on language, as in the case of writers, then you're in a bit of a fix. How can you acquire new landscape features in a zone not readily accessible from the rest of your landscape? You're going to need all the help you can get to shake everything together. Mood swings, particularly manic ones (which are often associated with depressive episodes too), are one source of aid; booze is another. There's another tentative prediction lurking here. Novelists who show greater EEG coherence between speech centres and other brain areas (a likely measure of the degree of integration of speech landscapes with the rest) should be less prone to excessive drinking — at least while creating their works. Again, I don't know of any relevant research in this area. It's not the sort of thing likely to attract 'on spec' interest or research.

I suppose the overall message here is that, if you have a stable temperament and want to be creative, don't hope for a settled life — arrange to have someone invade your country, go abroad, switch careers in mid-stream, that sort of thing. It's ironic that one of the arguments offered for abolishing military National Service, which was very good at shaking people up, was that it interfered with the smooth progress of education. Unfortunately smooth progress, it appears, may be good for creating Jobsworths, but not for promoting creativity. It's lucky that students, perhaps recognizing this, so often insist on taking gap years, doing VSO or whatever — National Service equivalents in many respects, though usually a good deal more enjoyable.

Some of David Healy's innovators emphasized the role of luck in their achievements, too. In particular they thought it crucial that they had happened to 'fall on their feet' into environments where they could do their own thing — play, in other words — in the company of congenial colleagues. The huge productivity of the Cavendish laboratory in the 1930s and the Bell and IBM research institutes later in the twentieth century, where this sort of atmosphere was fostered, strongly suggests that they were right. Of course it's an atmosphere which is becoming ever harder to find these days, squeezed out by managerial concepts of 'targets,' 'productivity,' 'accountability' and the like.

Social dynamics at work

The next topic has to do with the fact that attractor landscapes tend to become ever more rigid and immutable the more they are used. Basins deepen and widen, connecting routes get worn smooth and hard to alter. We saw in Chapter 3 how evolution has been forced to adopt an extremely risky strategy, namely sleep, to mitigate this problem in the case of Levels 1 and 2 dynamics. I want to take a look at some implications for social network dynamics in which, if the model has validity, it must also cause major problems. Here's an example of the sort of thing I have in mind.

Psychiatry in the UK was hugely productive and innovative over the twenty or so years following World War Two. Group therapies, therapeutic communities, day hospitals, industrial therapy, the idea of 'institutionalization' and much more, were all of mainly UK origin, while the whole subject was also greatly stimulated by psycho-pharmacological innovations coming from abroad. Spokesmen (they were nearly all male then) for psychiatry like Sir Aubrey Lewis of London's Maudsley Hospital had a prestige that modern equivalents can only dream about. But the professional organization was chaotic. There were centres of excellence which really were excellent, and snake pits; highly talented staff, and those who should never have been employed.

Psychiatrists were loosely affiliated with the College of Physicians, and their main qualifying examination was something called the 'DPM.' This was not a difficult qualification to get, and could be obtained about two years after first starting in the field (only those already possessing medical qualifications were eligible). In London at least an important requirement for passing was physical fitness, since the examination was held on the top floor of a building in which the lifts were unreliable at best. So you had to walk up the stairs. Some older candidates would arrive at the top blue, gasping and clearly incapable of writing anything. However, provided people were able to cope with the stairs and had done a fairly minimal amount of reading in the subject, they were pretty sure to pass. You might well think that this was a terrible way to select psychiatrists — let's see if it was.

Along with a whole range of other specialists — pathologists, gynaecologists and so on — UK psychiatrists started to clamour for their own college, independent of the physicians, in the late 1950s. They finally got it in 1971. The college set to work energetically and busily doing the sorts of thing that medical colleges do; raising standards, promoting education, conducting rigorous examinations and so on and so forth. All good, worthy stuff that everyone could see was worthwhile.

And the state of British psychiatry after all this effort? Well, some things are improved. The big snake pits have gone, though a host of new, smaller ones have appeared in the form of some of the less satisfactory 'community care' facilities (more usually lack of facilities!). Education standards in the profession have, on average, risen. However, no one could accuse UK psychiatry of being innovative these days. There is no longer a lot that is absolutely awful about it, but there isn't much that's outstandingly good either. The old, far looser, system for educating and selecting psychiatrists certainly had its downside, but also appears in some respects to have been better than the current one. Psychiatrists themselves mostly blame social trends outside their control, and government under-funding, for their predicament. All the same one wonders why it is that a more numerous workforce, apparently well trained and qualified compared to the pre-1971 generation, should be achieving less. Psychiatry isn't alone here, of course. Lee Smolin has given a beautiful description recently of how a very similar problem is afflicting the theoretical physics community.

Maybe this isn't so mysterious after all. The College of Psychiatrists has all the hallmarks of a bureaucracy, and everyone knows the effects that these can have on any field. Pulling in ever greater resources, often staffed by enthusiastic, well-meaning and capable people, these organizations spawn committees and have an increasingly stultifying effect on all who come under their sway. What's going on is that the attractors of bureaucratic dynamics become steadily less fluid as the bureaucracy matures, with continually deepening basins of attraction. This petrifying landscape puts increasing constraints on the Level 2 dynamics of everyone who comes into contact with it, thus stifling individual innovation — and effectiveness too if the bureaucratic landscape is sub-optimal in relation to particular functions, which it usually is.

The people involved are generally aware that there are problems and have two stock solutions for them. The first is more of the same; more intensive educational efforts, more frequent or rigorous examinations, and so on. The second is to re-organize. Both 'solutions' almost inevitably make the problems worse. More of the same simply deepens the attractor basins that caused the difficulties in the first place. Re-organization generally has much the same effect, for it is always imposed from above rather than emerging from below. So it leaves existing attractors largely untouched, except that their basins often become more extensive as the re-organization proceeds.

What solutions *would* work? Taking a leaf from nature's book, what's needed are destabilization and sleep equivalents. Destabilization is the most risky — too much of it and you end up with anarchy and mayhem as in the earlier stages of the French Revolution or Chairman Mao's cultural revolution. But lesser amounts ought to be effective. To go back to the College of Psychiatrists example, measures like randomly altering a proportion of the educational requirements and examination syllabus every year or two could be useful. It could be announced in advance that next year people would be examined in art history, say, or ecology, instead of social psychiatry or cognitive theory. Perhaps unfortunately for this proposal, it would be necessary to keep a few core subjects always there — such as psychopharmacology, if only to reduce the risk of future psychiatrists poisoning their patients. Then again, maybe a proportion of committee positions should be allocated to people selected at random from a large pool, like jury service. This should apply particularly to senior posts. After all there is a lot of truth in the aphorism that anyone who wants to be a president/prime minister is unfit for the job.

The sleep equivalent option is more interesting in a way, because it's not obvious why it should work. However, it is one of nature's favourite solutions and clearly *does* work in relation to the dynamics of our brains, so may well be effective in social dynamics also. What form could it take? There are lots of possibilities, but the most straightforward might prove most useful. All committees belonging to a bureaucracy should simultaneously have regular prolonged periods, say one month in every three or three months in every year, when they continue to meet as usual but are deprived of all agendas, minutes and secretarial help. They could occupy the time as they saw fit (provided they did not agree in advance on how to occupy it), watching television together, reading out loud to one another, playing parlour games, or whatever. The prediction is that, rather like Adam Smith's 'invisible hand' emerging from the market to guide the market, flexibility and fluidity would emerge from a procedure of this sort.

These suggestions are fairly unrealistic, if only because bureaucracies like the College are mere dimples in a much larger social landscape whose form severely limits what they can do. Imagine what the press would make of it if they were to discover that psychiatrists were being trained as art experts, for instance, or that there was nobody available at headquarters to answer queries because they were all 'asleep.' But there are related predictions to be made that are not in the realms of fantasy.

The largest social dynamic in the world today is the one involving the internet. It has not yet formed a rigid attractor landscape of its own because it has hitherto been continually destabilized by rapid expansion and technical advance. However fixed landscapes will form sooner or later, I'd guess about ten years down the line. People will begin to experience increasing constraints on what they can do when using the internet, whose exact nature will depend on the shape of the newly rigidifying landscape. Users won't be fully aware at first of what's going on, or of how they are being shunted into pre-determined channels. But it will eventually become obvious and will be viewed as problematic. There are straws in the wind already. For instance, since I changed email address, one small American college won't accept my mails. Yahoo says its server 'doesn't like' them. It's the only academic institution out of more than fifty in my address book that has this aversion and there is no obvious reason for it. Efforts (by the man I'd like to mail) to get the server to change its mind have failed. My contacts are being moulded, not by any technical failure or limitation as used to be the case, but by the internet server's expressed preference. My prediction is that non-understandable constraints apparently imposed *by* the internet, without the involvement of human intention, will become ever more frequent and pervasive.

Numerous solutions to the perceived problems will be suggested and none of them will work. The situation will go on deteriorating and may get sufficiently bad for more radical suggestions to be made. Two of these *will* work. One would be to remove all intrinsic memory from the system — all those cookies and favourite place records and so on — thus destabilizing it. The other would be to shut it down entirely (except for essential functions) for regular periods whose optimal frequency and duration would have to be determined by trial and error. If things had got really bad, both measures might be needed.

We're now in a position to answer the two questions posed at the beginning of the chapter (that is, what use is the attractor dynamics model and how can it be tested?). Its main use is to provide explanations of a whole range of phenomena from the functions of sleep (Chapter 3) to the evolution of bureaucracies. The explanations in turn suggest particular interventions which may prove helpful in diagnosing and treating some medical conditions (Chapters 3 and 9) or, as we've seen in this chapter, in promoting creativity or easing the pangs of bureaucracy. And these are only a few among a great many suggestions that could be made.

Experimental tests of small parts of the model are possible. Freeman's work on rabbits (Chapter 1) can be regarded as a test of the concept of Level 1 dynamics. The prediction that emigrant geniuses should be less prone on average to affective disorders than stay-at-homes would test an aspect of Level 2 dynamics. The studies showing that milk drinkers have unusual genetics can be viewed as a rather indirect test of the proposed linkage between social and genetic dynamics. But the uses themselves are what could provide the broader tests that are needed. If they turn out to be useful in real life, then confidence in the whole model will be reinforced. If they don't prove effective, the model will need to be modified or abandoned. Time will tell!

15. Entanglements

Paul Marshall concluded, towards the end of his book on extrovertive experience (see Chapter 12): 'Physics was the route by which mind was excluded from conceptions of the world at large; and physics may be the route by which mind finds its way back in.' Lots of people have recently been seeking this particular 'route back in.' Karl Popper, the philosopher famous for introducing the idea that 'refutability' is what differentiates science from non-science, and Nobel Prize winning neurophysiologist Sir John Eccles, were among the pioneers here. They published a book in 1977 *(The Self and its Brain)*, in which they argued that conscious mentality is down to a manifestation of quantum events in synapses. More recently, the brilliant mathematical physicist Sir Roger Penrose caused a stir with his *The Emperor's New Mind*, published in 1989. In it, he proposed that consciousness is a function of quantum collapse in the brain. He went on, in collaboration with American anaesthetist (anaesthesiologist) Stuart Hameroff, to develop a detailed theory about how quantum coherence in intra-neuronal structures known as microtubules underpins consciousness. Many others have followed similar paths. A plethora of such theories were proposed towards the end of the last century.

Many of these proposals, including the Penrose/Hameroff 'Orchestrated Objective Reduction' theory, depended on the occurrence of quantum coherence between fairly large numbers of separate particles in the brain. I had better try to explain what 'coherence' is. Mathematically, it's a state in which numbers of previously separate particles come to share a *single* (Schrödinger) wave function and thus become unified themselves. Physically, it is what is responsible for phenomena such as super-conductivity when materials lose all electrical resistance, and super-fluidity when liquids lose all resistance to flowing and can do bizarre things like pouring themselves out of upright containers. So-called 'Bose-Einstein condensates' are coherent, and there was a good deal of excitement recently when advances in low temperature technology led to people being able to create some of these out of formerly separate metal atoms. One such condensate contained about sixty ruthenium atoms. In terms of what we've discussed in this book, condensates are systems like an Alice and Bob pair of entangled particles, except that they are more extensive — not necessarily spatially extensive, but extensive in the sense that they involve larger numbers of particles.

Therein lies the problem for proposals relating coherence to consciousness. For the whole environment, including all the thermal radiation that permeates it, can act like Eve in relation to coherent states and cause rapid decoherence. To get metal atoms to form a Bose-Einstein condensate, for instance, you have to cool them to within a whisker of absolute zero using amazingly sophisticated laser techniques — there's no way you could make one in your home freezer. It has recently become possible to calculate how fast decoherence would happen to a state in the brain that started off coherent. The answer is that it would occur in less than a millionth of a billionth of a second for most coherent states. There are a few such states that are less vulnerable. What's called 'spin coherence' between atoms in cell membranes is a bit less susceptible to the 'Eve effect' according to calculation (an unpublished calculation by Erhard Bieberich), but even so decoherence would occur in around a thousandth of a billionth of a second. This seems far too rapid to allow consciousness a look in. That operates on time scales of the order of a tenth of a second at fastest.

My guess is that many of these theoreticians have been looking in the wrong direction. They've been hoping to explain consciousness in terms of quantum theory, whereas its actual relevance may be to unconscious mind. Consciousness, I've argued here, is a mainly 'Level 2' brain function, and there is no obvious reason for invoking quantum theory in this connection. Reasons that were commonly given for doing so at one time (to do with the unity of consciousness and the so called 'binding problem') are already accounted for in the attractor dynamics picture, and no longer provide grounds for invoking quantum effects.

Mind, however, is mostly unconscious — or at least most of it does not possess what we would normally think of as 'consciousness' — and is a product of the hierarchy as a whole. Quantum theory is very relevant to the top of the hierarchy according to this picture. Some physicists have themselves proposed more general pictures, not unlike the one I am advocating; for instance, Amit Goswami, a professor of physics based in Oregon whose main book on the subject *(The Self-Aware Universe)* appeared in 1993. Goswami advocates a form of Idealism (the view that the material world is in some sense secondary to the mental world). I disagree with him there, my own view being that both mental and material worlds are manifestations of some deeper, inaccessible reality — what is sometimes termed the *unus mundus* in the German literature — but that's by the way. His overall idea that the universe as a whole is mind-like fits the view proposed here: namely that attractor dynamics play an essential role in mind at all levels, while the attractor dynamics at the top level are based on entangled entities that can only be conceived in terms of quantum theory.

All this is by way of preamble to thinking about how links between entanglement dynamics and the lower levels of the hierarchy could work. As pointed out earlier (Chapter 13), upward links give no great problem of principle. Everything that we think and do creates new entanglement relationships between some of the particles involved in our thoughts and actions, which will feed into the great sea of entanglement and affect what would appear to us to be its 'dynamics' if we could but view them. The fact that the entanglements for which we are responsible are miniscule compared to those already existing does not imply that they will have no effect. For entanglement dynamics must be largely or entirely chaotic — hence most attractors in it will be 'strange' and 'butterfly effects' will occur. In theory, my thoughts about what I would like to eat for lunch today could cause an entanglement storm affecting Andromeda tomorrow ... or indeed the fate of the Roman Empire, or that of some wonderful space colony in a million years time. That's where the weirdness is most obvious — in the lack of temporal constraints on the effects that upward links can be envisaged to produce.

But are there any downward links? Could entanglement storms in fact affect the real world? Because any such effects would neither be constrained by relativity nor involve ordinary physical causation, they could probably never be 'seen' in a completely unambiguous way even if they do occur. They could always be viewed as understandable in principle if not in practice, as outcomes of natural law or of chance, by anyone sufficiently determined to maintain a reductionist mind-set. Therefore, some would say, the whole topic must be regarded as an unprofitable speculation wholly outside the remit of science. As you'd expect, I disagree with any such view because, if entanglement effects do occur, there's a possibility that they may be detectable in relation to both our minds and to obscure statistical effects on other systems. Let's see how.

As the 'quantum consciousness' theorists were fond of pointing out, conscious minds can be affected by single quantum events. The arrival of a single photon of light at the retina of our eye, a quantum event *par excellence,* can under the right circumstances be perceived consciously. Because brain dynamics are 'poised on the edge of chaos,' there's no problem about envisaging how single quantum events within the brain could also produce large effects. Butterfly effects may abound. When the wave function of some electron, for instance, collapses, there is an apparently random element to precisely where it will materialize. Maybe, if it happens to materialize on one side of a nerve cell membrane, it will trigger a cascade of events that will lead to the owner of the membrane discovering the Theory of Everything; if it materializes on the other side, nothing happens and we never get the theory. Tough![1]

There's been lots of speculation about where the apparent randomness inherent in wave function collapse comes from. Some say that it is true randomness of a sort that doesn't seem to occur anywhere else, but the problem here is that there's no obvious home in nature for 'ideal' randomness. The so-called random numbers generated by computer programs are pseudo-random.[2] The 'randomness' of chaos is in fact no more than unpredictability; chaotic events are fully deterministic, it's just that we can't tell how they will turn out. The sequence of digits in irrational numbers like 'pi' or the 'golden ratio' is apparently random, but is nevertheless fully determined by the calculations that generate the numbers.

It seems reasonable to suppose that the apparent randomness involved in wave function collapse is like chaotic 'randomness;' that is, determined by factors so complex and so sensitively dependent on initial conditions that we have no hope of ever predicting them. After all chaotic events are usually fully predictable in a statistical sense, and so are quantum ones. Given large enough numbers of chaotic or quantum events you can tell what will happen on average, to a degree of accuracy which will steadily increase as you take note of ever more events. It does very much look as though these two types of apparent randomness share similar origins, in which case the outcomes of wave function collapse may be just as deterministic as those of classical chaos. Einstein at least would have liked the idea; one of his best-known assertions was that 'God does not play dice.'

And the source of the chaos producing quantum 'randomness'? Entanglement dynamics is ideally suited to the job. If it is indeed responsible, then the outcomes of wave function collapse could be regarded as fully determined in principle by the network of entanglements affecting the particle in question — which would depend on its entire history and that of whatever particles or apparatus it was interacting with at the time of its wave function collapse. This entanglement history would usually encompass a largish part of the universal entanglement 'sea,' so there would be absolutely no hope of tracing significant individual relationships within it, except in the case of 'Alice and Bob' like situations where particular particles had been kept isolated from the 'sea.'

The relevance of all this is that large numbers of particles can (almost) always be expected to behave in a statistically predictable way when it comes to the outcome of wave function collapse. It would take a huge fluctuation in the underlying entanglement dynamics to produce any detectable deviation from the expected statistics. However, single particles could well be sensitive to smaller fluctuations. Let's go back to the scientist on the verge of producing a Theory of Everything. Precisely where the crucial electron will materialize can be envisaged as determined by the chaotic dynamics of its entanglement history in relation to that of the relevant part of the scientist's brain. But this entanglement dynamic could itself be influenced by some attractor, which might alter it sufficiently to give 'Hooray, I've got it' instead of 'Oh dear, maybe I'll get it one day,' which would have occurred in the absence of the attractor. That's how a downward link could work. It could affect apparently random quantum events in systems sensitive to such happenings like our brains. And of course many of these events would feed back to affect the entanglement dynamic. Discovering a Theory of Everything would result in all sorts of thoughts and actions that could never have occurred in its absence.

What about some evidence for all this? To start thinking about some of the evidence that may be available today it's necessary to make another assumption — I'm sorry to say, since we already have a tottering pile of assumptions. But this extra one shouldn't cause the pile to overbalance and, in any case, not all of the evidence depends on it. It is that entanglement-influenced, apparently random, collapses in the brain can translate into meaningful experiences in our consciousness. In one way this is no big deal, for our consciousness is dependent on our brain, which in turn is ultimately a product of innumerable wave function collapses. But in another way it is mysterious, for many of the sorts of experience that might provide evidence of entanglement dynamic effects must be envisaged as moulded by the content of the dynamic somehow — in other words, there must be information flow from the entanglement realm that gets translated into specific meanings in consciousness. I can't offer any clear account of how this could take place via the proposed mechanism of 'deterministic collapse,' and can only speculate that it might be a consequence of upward feedback effects, which are not subject to temporal constraints once they get into the entanglement realm.

The more dramatic types of experience that might be regarded as evidential are anecdotal. There are certainly plenty of them. Common examples are reports of someone having a vision of a loved one at the moment of their death on the other side of the world. That sort of experience is easily dismissed by critics — unless it happens to them personally. Experiences like that imply information flow from some incorporeal realm into consciousness. A relevant anecdote that doesn't require this sort of information flow involved Jung's friend, the great theoretical physicist Wolfgang Pauli, who was renowned for the 'Pauli effect.' Whenever he walked into a laboratory where an experiment was running, the apparatus tended mysteriously to fail. Some experimentalists banned him from their labs in consequence.

Parapsychologists, of course, have tried to move on from anecdote to carefully conducted experiments on 'telepathy,' 'clairvoyance,' 'psychokinesis' and the like. Indeed the methodology that many of them use nowadays is a good deal more rigorous than that employed in most mainstream science.

What generally seems to occur is that, when some new parapsychology experiment gets off the ground, it gives strongly positive evidence at first for whatever aspect of 'psi' it is supposed to measure. Then the significance of results falls off and negative findings begin to surface. Parapsychologists themselves have noticed this pattern and claim that positive results often return if experimenters persist for long enough. Persistence, however, is quite rare if only because most experiments are so rigidly designed and controlled that they are extremely boring to all participants. After a few days guessing cards in a sound-proof room in a telepathy experiment, for instance, many people would rather have a hole in the head than ever see another card.

Some parapsychologists suggest that it is the boredom which causes the fall-off in positive results. They're caught in a cleft stick because, if they try to enliven their experiments, they are then open to accusations that any positive results are due to 'poor control.' But boredom can't be the whole story because even people coming new, fresh and enthusiastic to some particular experiment often produce fewer positive findings than did the pioneers. This isn't the place to go into detail, but I should add that parapsychologists have successfully excluded all obvious, mundane reasons for such patterns (for instance, earlier experiments being based on worse methodology; negative results not being reported at first, and so on).

Critics have no great problem in putting results and alleged findings of any individual experiment down to chance fluctuations, by which they mean that the findings have no meaning. I suggest that the critics are quite right to put them down to fluctuations of chance, but that they're right for completely the wrong reason — because it is fluctuation in 'chance' that *produces* 'psi'! And, in fact, the overall evidence that weak 'psi' effects do occur is extremely strong. In statistical terms, it is much stronger than the evidence that antidepressant drugs help to cure depression, for example.

If 'parapsychology' is mediated by the entanglement 'sea,' then 'psi' effects must be attributed to what can be envisaged as waves in the ocean. And waves, of course, pass — at least they do from our point of view, even though they can be regarded as 'frozen' from the 'block universe' (Chapter 13) point of view. So the pattern that parapsychologists have noted, of changing statistical outcomes, is just what would be expected. Have such waves ever been visualized? What we're looking for is wave patterns with an overall resemblance to the EEG ones that led Walter Freeman (Chapter 1) to propose that chaotic attractor dynamics are at the basis of perception. Unlike EEG however, any 'entanglement waves' won't be pen movements recording electromagnetic field changes, but will manifest in regular fluctuations of measures of 'chance' events. Moreover there's no reason to expect their typical time scales to be anything like that of EEG (that is, approximately 1–100 cycles per second).

Dean Radin is a leading American parapsychologist who has written a number of reviews of the entire field. There are two particularly interesting (from my point of view) graphs in his book *Entangled Minds*. Both show results relating to 'chance' events over a period of time. Radin recorded this data in order to give a background for 'blips' in the data relating to particular brief events — in one case the 9/11 catastrophe (Radin was recording his data expecting that that *some* catastrophe would occur if he continued recording for long enough, and 9/11 was the catastrophe that *did* occur). The other brief event was a 'healing' session, whose timing was known in advance. He wasn't concerned, in other words, with the background itself except in so far as it provided a contrast to what occurred during the events. But, if one *does* look at the background, it appears remarkably like waking EEG activity. Radin shows fluctuations in performance on an ongoing card-guessing test, when people were asked (online) to predict which of five different types of card would later be 'randomly' selected by a computer (illustrated in

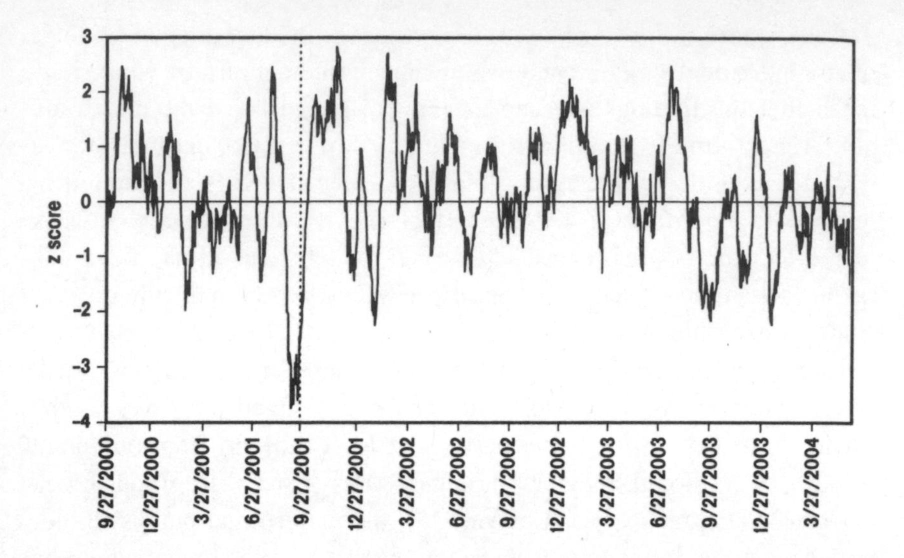

Figure 7: Fluctuations in online ESP card test performance on a daily basis from September 2000 through June 2004.[3]

Figure 7). This was done daily over a three and a half year period. There's a dominant frequency of approximately 1 cycle/ 3 months, which looks very like an EEG alpha rhythm (~10cycles/second). As in the EEG case, there are also whole lot of other, lower amplitude but mostly higher frequency, frequencies superposed on the dominant one. Incidentally, there was a 'blip' associated with 9/11. Peoples' guesses were quite a bit *less* accurate than usual over the preceding day or two.

In another study, Radin shows the outputs of three separate random number generators over a ten-day period, combined to give a measure of their 'coherence' (illustrated in Figure 8). The output of two of these generators depended on quantum level electronic happenings, while that of the other was due to random ionizing events — that is, the numbers were thought to be 'truly' random; they were not generated by a pseudo-random computer programme. The resultant picture of the degree of coherence between these supposedly random numbers was again EEG-like, the dominant frequency having a cycle time of about twenty hours. So outcomes of both (supposedly chance) card-guessing and random number generation appear to be associated with EEG-like wave pictures. This is not proof that a largely chaotic entanglement dynamic exists capable of affecting 'random' events, but is an indication that it may exist.

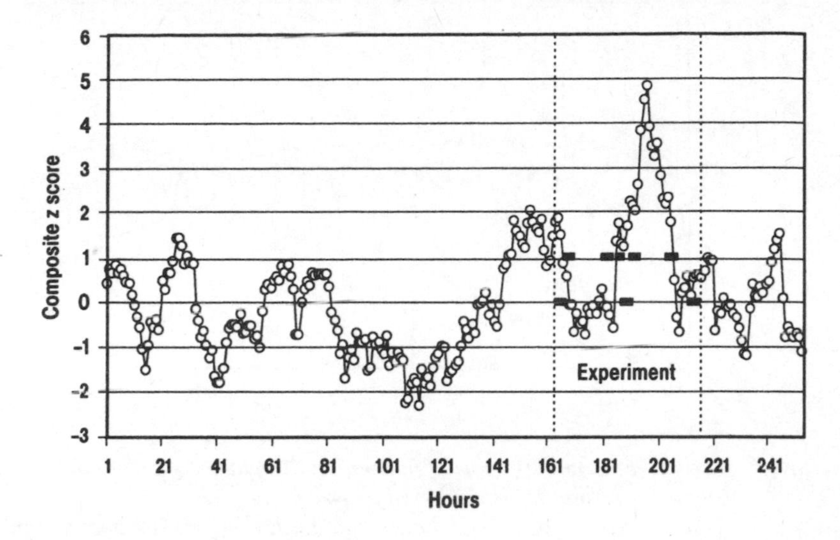

Figure 8: Combined results of three RNGs.[4]

To introduce a note of caution, I should point out that, because Radin's plots look like EEGs to the naked eye, doesn't mean they are like EEGs mathematically. Plots of truly random walks can look like EEGs. Given the different time scales involved (~10 cycles per second for EEGs, versus 1 cycle per day or per several months for Radin's 'chance' cycles), it would take many years of recording and effort to prove any similarity with EEGs. All the same, the resemblances are suggestive. A more direct test of the idea that an 'entanglement dynamic' affects 'random' events might involve looking for regularities in fluctuations of rates of radioactive decay, for instance, which are usually regarded as strictly random, quantum happenings. One practical difficulty here would be to make an apparatus sufficiently 'tuned' to the (unknown) frequencies involved to be capable of detecting them. There's a statistical problem involved because, if you made the apparatus sensitive to an infinite range of frequencies, it would be bound to come up with lots (in fact theoretically an infinite number!) of actually meaningless, but apparently positive, findings. On the other hand, almost all such spurious 'positives' would be very short-lived. So, with time and persistence, it should be possible to home in on any genuine 'entanglement dynamic' frequencies.

Blue skies research does sometimes prove worthwhile, though this particular suggestion would take a lot of effort to actually do. It would be

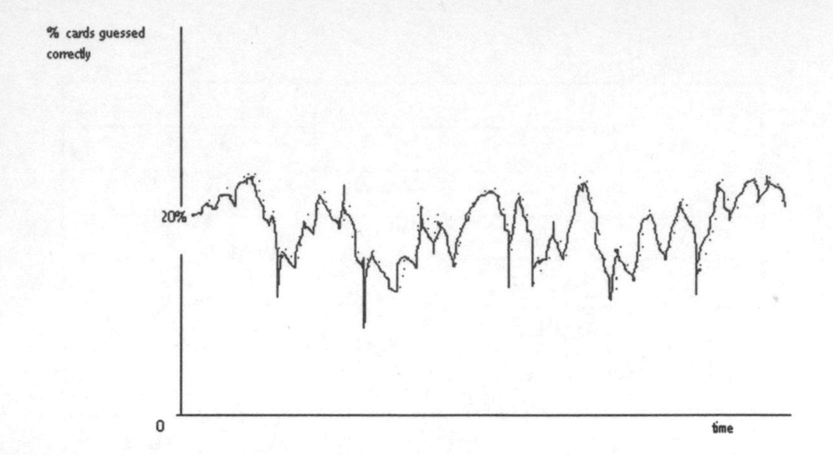

Figure 9: A representation of the sort of fluctuation of chance seen by Radin. Unlike his result, this is not based on real data.

The imagined experiment here is that large numbers are guessing in advance which of five different sorts of card will be randomly shown at a particular time. On average, they should guess right 20% of the time. However, as the graph shows, the 'actual' percentage of correct guesses fluctuates quite a bit.

The graph resembles an EEG recording. If it were based on real data, would it be mathematically similar to an EEG? The answer to this question is unknown, but is arguably worth researching.

looking for EEG-like order in supposedly random quantum happenings. But these happenings *are* random on most time scales. Any order could be expected to show up, so Dean Radin's plots suggest, only in relation to very low frequencies, and would thus need records extending over many years for detection.

Looking for wave-like behaviour of 'chance' aside, what other sources of evidence might there be for the existence of an entanglement dynamic? Systematic investigation of the types of drug experience described in Chapter 4 could well turn up material of interest. It will be recalled that the 'black pumas' seen by European ayahuasca imbibers looked more likely to be denizens of a top of the hierarchy realm than of the bottom, genetic level. And there were good reasons for supposing that they didn't have their origin in the social dynamics either, which leaves only the entanglement realm as a home for them. They kind of popped out from Benny Shanon's enquiry into the phenomenology of ayahuasca experience, which he undertook for reasons that had nothing to do with trying to find out whether an entanglement realm has effects on us. It's intriguing to speculate on what might be found if one were to look specifically for entanglement effects. All sorts of paranormal experiences are claimed by drug takers, which beg for careful investigation. If only one in a hundred of the claims, of 'visiting' foreign countries for instance or of 'witnessing' historical events, were found to have some sort of validity, this would be of enormous interest and might tell us a good deal about entanglement dynamics and their consequences for our conscious experience.

Quite a bit of research, some of it directed towards questions of this sort and mostly using LSD or sometimes mescaline (another hallucinogen), did start up in the 1960s. A range of fascinating findings were just beginning to emerge when all such research fell victim to friendly fire at the onset of the 'war on drugs.' Some of the research was of practical interest; for example there were claims that LSD was useful in the treatment of personality disorders and addictions, both of which are still mainly resistant to today's treatments. Other research had to do with what is now called 'transpersonal psychology' — particularly claims about various 'psi' phenomena.

The war on drugs continues unabated and it remains difficult in consequence to research hallucinogens. However the situation may change since ever more people are beginning to realize that it (that is, the 'war') has been amazingly effective at creating and enriching criminals, but does not appear to have curbed increasing drug usage. Maybe usage would be even higher without the war. Or maybe people would have taken more responsibility for their own behaviour if there had been less state intervention, and would have cut down their usage. Who knows?

In the meantime, there are alternative research strategies that are both worth pursuing and are being pursued (see Dean Radin's books for clear, reliable accounts). Quantum effects on the content of consciousness are likely to manifest most readily when attention is switched off — for attention keeps one focussed on the here and now. Therefore dreams and meditative states where the goal is attentive shut-down, for instance, merit further study. Systematic collection of information about unusual experiences and their timing could be of interest. There's a good deal of information about allegedly clairvoyant dreams and the like, and a certain amount relating to apparent 'psi' in meditators. However none of it is critic-proof. On the other hand, there appears to have been little systematic work on any inherent phasic changes in these phenomena — the prediction being that they should fluctuate in incidence in phase with an underlying entanglement dynamic. Most of the research tries to relate fluctuations to known 'measurables' of some sort. The main practical problem here is that, if fluctuations were to be found, it could prove hard to unambiguously distinguish ones due to the social, memic dynamic from any due to an entanglement dynamic.

The so-called 'ganzfeld' technique attracted a lot of interest a decade or two ago. Basically, it involved putting half ping-pong balls over people's eyes, so they could see only white light (in many experiments it was in fact pink light, as the ping-pong balls had been coloured pink), and playing formless noise (white sound or whooshing noises, for instance) into their ears. It was a somewhat half-hearted variation on an earlier technique of inducing total sensory deprivation by having people lying in the dark in warm swimming baths, while wearing ear-plugs. The ganzfeld experimenters may have been cautious because there were reports that total sensory deprivation can occasionally cause long-lasting psychotic states. Anyhow, it was claimed that ganzfielders were good at feats like telepathy and remote viewing. The CIA and the US military both, it appears, ran programs to investigate these claims. However, as tends to happen with 'psi' claims generally, initial strongly positive findings apparently soon declined so much as to discourage continued interest in practical applications, such as clairvoyant spying. Someone, somewhere, holds a lot of data on all the research that was carried out (over a 24 year period, it is said). If phasic patterns do occur, evidence for them is more likely to be buried in these findings than in most. It might well be worth a look-see.

But perhaps I've said enough to suggest that, if an 'entanglement dynamics' realm exists, it is not necessarily going to prove forever inaccessible to science and the probings of the truth meme.

Just a final thought before I close. You've probably noticed that I've discussed beauty and truth in this book but have kept quiet about love, the greatest of that trinity. We decided that beauty has its home in Level 2 dynamics, while truth is principally lodged in the memic world of social dynamics. I suggest it makes sense to suppose that love has its main base in the entanglement realm. There's a nice symmetry to the idea, plus there's no obvious basis for love, as opposed to lust, enlightened self-interest and the like, in the other levels. If you want to get an inkling of what that realm may truly be like, therefore, forget all the intellectual stuff in this chapter. Instead, think charity and universal compassion; think mothers cuddling their babies; think of the best sex you ever had, when you could no longer tell who was you and who was your lover ... That may offer a glimmer of insight into what entanglement, and the universe, is actually about.

Addendum

Ongoing discussions, especially with Peter Henningsen, are making it ever more clear that a crucial factor affecting the picture outlined in this book has to do with defining the state spaces in which attractors are occurring, and the boundaries of the concomitant landscapes. For mathematical purists, attractors can be regarded as occurring only in 'closed' systems, that is, systems unaffected by outside influences. The only totally closed system is the universe as a whole. Therefore, if one is a purist, one should talk only of equally universal attractors. However, as all the work on neural networks for example shows, it can be useful to think of attractors in systems which are to some extent open. From a practical point of view, it seems that a system can harbour effective attractors provided the strength of outside influences on it does not exceed some threshold.

There are two ways of thinking about what happens when outside influences do exceed the threshold. One is to say simply that they may often disrupt or suppress the attractors in the system. The other, which I believe is more fruitful, is to say that the system becomes incorporated into a larger one, which harbours its own attractors, and that these new attractors 'take over' from the original ones in defining the behaviour of the components of the original system. Of course, some of the new attractor landscapes might be enlargements of, or at any rate similar to, ones in the original system, but others may be dissimilar. So, when outside influences exceed a threshold, the state space of a system will lose its own identity in contributing dimensions to a larger state space — that is, one that describes both the original system and the influences on it — whether or not the overall shape of the new landscape differs from the old.

This might seem a rather arcane point were it not for the views of Max Velmans. He is a psychologist who, for at least fifteen years, has been plugging the proposal that conscious perception is best thought of as some sort of amalgam between goings on in the brain and happenings in the world beyond. The mere fact that we experience what we see or touch as being out there in the world or in our finger tips, and not in our heads, leads to this conclusion he argues, calling his theory 'reflexive monism.' For many of us, me included, it seemed a very cloud castleish proposal for there appeared to be nothing on which to base any 'monism' — after all, the visual cortex of your brain is one thing and the tree that you're looking at over there is quite another, which implies a dualism not a monism. However, if influences of the visual environment on the brain do exceed a threshold, it must be entirely legitimate to regard the situation monistically in the way Velmans proposes, since brain and relevant aspects of the environment must be regarded as occupying the *same* state space when it comes to defining where the appropriate attractors 'live.' And of course the mere fact that we perceive something visually can be taken to suggest that the threshold *has* been exceeded.

Max himself (personal communication) points out that everything has to be regarded as simply a chunk out of the universal whole. And therein lies both an enormous difficulty and a major opportunity. The greatest difficulty arises from the fact that the chunks of the universal dynamic described by particular attractor landscapes must be assigned *temporal* boundaries as well as spatial ones. This is not too much of a problem when considering simple systems like the weight on an elastic band example of Chapter 2. The spatial boundaries of factors contributing to the relevant landscape are clearly visible, while their temporal extent is the period between first starting to move and arriving at the point attractor. But it's far harder to define what should be included in the state space relevant to me looking at a tree. The spatial boundaries separating what should be included from what should be excluded are fuzzy at best, but the biggest problem is to do with the appropriate temporal boundaries, for memory is known to be essential to perception — before we can see anything properly, we have to *learn* how to see it — so the dynamics of my perception of a tree today will be affected by the dynamics of all my past experiences of trees right back to infancy. This implies that factors contributing to the relevant attractor landscape would have to include a large chunk of my history as well as what's present in front of my eyes. That's the difficulty, and it's a huge one from a pragmatic point of view for there's no easy way of ascertaining which bits of history should be included, and which could safely be left out.

The opportunity is to do with the fact that consciousness is known to be tied up with early stages of the memory process in some ill-understood way (see my *De la Mettrie's Ghost* for an account of this). Perhaps a proper understanding of consciousness is lurking somewhere within considerations to do with the temporal 'stretching' needed to define landscapes relevant to some brain dynamics. I attended a conference on 'Mind and Matter' in the summer of 2006, where a number of participants suggested that a proper understanding of conscious mind would one day turn out to be linked to an understanding of temporality. Maybe we are beginning to get an inkling of *how* this might be the case.

It's not only perception to which considerations of this sort may apply. Phenomena like mass hysteria, or the social 'attractors' I've described in the book, may also be best regarded monistically, the brain processes of individual participants being literally subsumed into larger wholes. There's already material for many more books here, and we are only just starting to scratch the surface of all the possibilities. Readers, please watch this space.

Notes

Chapter 1

1. His swings of opinion are illustrated by the following quotations: ' ... in its basic structure the human psyche is as little personalistic as the body. It is far rather something inherited and universal. The logic of the intellect ... the emotions, the instincts, the basic images and forms of the imagination, have in a way more resemblance to Plato's *eida* (sic) than to the ... whims and tricks of our personal minds.' (1939). This later changed to: 'Nobody would assume that the biological pattern is a philosophical assumption like a Platonic idea or a Gnostic hypostasis. The same is true of the archetype. Its autonomy is an observable fact and not a philosophical hypostasis.' (15/2/54). Despite what he wrote in 1939 which could reasonably be taken to imply that he considered archetypes (that is, the 'basic images and forms of the imagination') to be genetically inherited, eight years later he stated, though admittedly in the course of a quarrel: ' ... you are utterly mistaken in saying that I have described the archetypes as given with the brain structure.' (5/8/47).
2. The technique called MEG (magnetoencephalography) is much more recent than EEG since the ability to measure tiny magnetic field changes, like those produced by the brain, did not emerge till the 1970s. As one might expect, it's a sort of mirror image of EEG in that it is mainly responsive to current flows along fibre bundles, which contribute only a small component to EEG change.
3. By permission of Walter J. Freeman and C.A. Skarda. Skarda, C.A., Freeman, W.J. (1987) 'How brains make chaos in order to make sense of the world.' *Behavioral and Brain Sciences* 10: 161–95.
4. There's still debate over whether waking EEGs are ever fractal in the technical sense of being self-similar over a wide range of scales and possessing a single, well-defined Hausdorf dimension. The answer is probably that occasionally they are, sometimes they definitely aren't, and mostly they verge on being so but don't quite make it as somewhat different structures appear on different scales.

Chapter 2

1. If all the Lyapunov exponents relating to an attractor are less than 0 it will be a 'point' attractor; if one of them is 0 and others less than 0, it will be 'periodic'; if any are more than 0 it will be 'strange.' These exponents are measures of changes in ellipsoids.

2. Another way of looking at this is to say that, when a system becomes too 'open,' it loses 'system-hood.' In other words it becomes an integral part of a larger 'system' that also encompasses whatever is influencing the original one. The larger system will of course harbour its own, more general, attractors.
3. However, as Richard Feynman pointed out, 'there's plenty of room at the bottom.' What he meant was that there's as much 'space' between the scale on which subatomic particles have their existence and the very bottom level (the Planck scale) as there is between subatomic particles and the cells of which we are composed. Plenty of room for complexity in that space, therefore. Since cells are built from particles and the atoms into which they assemble, one cannot rule out the possibility that particles may have antecedents as complex as those of cells. Maybe electrons are emergent properties of such antecedents, as the string theorists appear to be saying, in which case they may not be so different from attractors as common sense would suggest
4. Some would say 'quantum field theory' rather than 'quantum mechanics.' I prefer the 'mechanics' appellation, even though quantum field theory is more up to date and general for most purposes. Field theory implies an infinite number of degrees of freedom, whereas quantum mechanics doesn't necessarily do so. Thus attractors would seem to have a more natural dependency on the latter, though this is a rather nit-picking point.
5. A theorem due to Emmy Noether, said by some to have been the greatest female mathematician of all time, who proved it during the First World War. It says: 'For every symmetry exhibited by a physical law, there is a corresponding quantity that is conserved.'

Chapter 3

1. There's pretty good evidence that long-term memory is to do with patterns and strength of connections in networks of neurons. Lots of connections get pruned in the course of development. New ones can form apparently as a consequence of learning. Dendritic spines, which form most connections, can wriggle about and change shape as readily as sea anemones. Changes in the effectiveness of connections are also known to occur. The one that has attracted most attention is 'Long Term Potentiation,' an increase in effectiveness associated with use shown by a particular type of connection (the NMDA subtype of glutamine receptor). Other types, including 'Long Term Depression' (that is, weakening of connection strength), are also known to occur, nor is the NMDA receptor the only sort able to alter its strength. The suggestion about changes in the strengths of interconnection being responsible for memory actually long precedes modern neural network theory, having been originally proposed by Donald Hebb in 1949.

2. As so often happens when ideas are 'in the air,' they surface in a number of different forms. For instance, after I'd completed most of this book, I came across Ben Goertzel's approach. He introduces many of the same notions as Henningsen, but on a basis of 'patternist' thinking. This is probably a better basis than Henningsen's in relation to Goertzel's central interest in Artificial Intelligence, but would have been confusing from my point of view, had I met it earlier, since it would have been much more difficult to apply to the central themes of later sections of this book, which are to do with social dynamics and entanglement. Similarly, Steve Lehar's proposals about 'Harmonic Resonance' offer an intuitive picture of what can be regarded as manifestations of attractor dynamics in the brain, but would be difficult to apply more widely.
3. The idea of Levels 1 and 2 should probably eventually be replaced by a concept of expanding or contracting state spaces. However this is more tentative, and harder to understand, than the 'levels' so I have stuck with them. They are quite adequate for the purposes of this book.
4. There's a very recent report (see Marshall *et al.* under Further Reading) that slow wave sleep helps memory for facts, in that artificially 'boosting' it also boosts overnight improvements in recall of word lists, but has no effect on performance of a 'finger-tapping task.' So far as I can see, these findings are consistent with the suggestions about sleep functions made here, though they cannot be regarded as providing good evidence, for or against, since they are open to many interpretations. Moreover I am personally sceptical of the validity of the finding since, although the number of words recalled increased significantly more after sleep in the group receiving a real boost than in those receiving a 'sham' boost, the two groups did not differ much in actual recall rates after sleep (41.27 words ±1.21, versus 39.50 ±0.84).

Chapter 4

1. There is anecdotal evidence about this from non-NDE experiences of autoscopy. The medieval Kabbalist, Yehuda ben-Nissam ibn-Malka, wrote of his own, self-induced autoscopy: 'Know that he sees nothing other than himself ... as one who sees himself in a mirror ...,' while there have been occasional reports of 'mirror-like' images of themselves seen by neurological patients (see the Arzy *et al.* reference under Further Reading). However, it's not clear whether these sources actually made a distinction between a self-image seem from another's point of view, or a self-image seen in a mirror.
2. It's not as though people *never* see big cats other than pumas or jaguars, either. Shanon reports very occasional visions of both tigers and cheetahs.

3. On this argument we should be even more likely to harbour 'hyena attractors,' since they are Africa's most fearsome predators in relation to medium-size animals like us. Although they scavenge if they get the chance, they are mainly very effective nocturnal predators. Lions, scavenging *their* kills sometimes get the 'credit' for having made them. But we don't obviously harbour any such attractor.

Chapter 5

1. A psychologist (Frederick Malmstrom) has recently suggested that the way 'greys' look reflects a newborn's 'recognition template' of its mother. Colour vision is thought to develop after birth, and mother's face seen from an infant's perspective might plausibly resemble a typical 'grey.' If that's the case, the social dynamic has served only to re-awaken a very ancient, inborn perceptual attractor, rather than helped to create a totally new one. Of course, if Malmstrom's suggestion is correct, there is then a problem explaining how other 'little people' came to be so often hairy and colourful.

Chapter 7

1. Douglas Hofstadter, whose book *Gödel, Escher, Bach* (1979) did so much to set the ball of consciousness studies rolling, preferred the term 'scheme' to 'memeplex.'

Chapter 11

1. Actually its immediate neighbour Scafell Pike is a few metres taller, but does not look nearly as impressive as Scafell from the usual (western) approach. So Scafell often gets the listing.
2. The golden ratio is the one that arises when you divide a straight line in two, in such a way that the proportion of the shorter segment to the longer is the same as that of the longer segment to the whole. It's an irrational number (~ 1.6180339887 ...).

Chapter 12

1. Not all such recurring discoveries are of the same frequency. Coming up with the claim that 'flying wings' are the aircraft of the future happens every twenty years or so. Publicity for flying cars and claims that we'll all soon be commuting by air may be of even lower frequency — around 25 or 30 years — perhaps because it's such a daft idea.

Chapter 13

1. One cannot say how Captain Kirk would perceive the timing of events while going at warp speed, for travelling that way involves new (and probably impossible!) physics that we know nothing about. If ordinary relativity rules applied, he would appear to us to be travelling backwards in time and would presumably himself perceive the Dalek/Sirian fleet movements as time reversed — that is, he should see them going from their destinations backwards to launch.
2. The weak processes violate what's technically known as 'parity conservation,' but the trio of 'parity,' 'charge' and 'time,' taken together, is conserved.
3. It's not at all clear, to me at least, whether a wave function collapse resulting in what's called a Bose-Einstein condensate would be associated with entropy increase, for these condensates are very highly ordered indeed. So far as I know their entropy has never been calculated. Maybe they are exceptions to the rule. They are very rare phenomena, so would not be expected to make any difference over all. But perhaps it might be interesting to look for time anomalies associated with them. Some types have already been shown capable of slowing light down to almost a snail's pace.

Chapter 15

1. Chris King, a mathematician based in New Zealand who has long had an interest in consciousness and cosmology, has provided a more detailed account of how this sort of quantum effect on the content of consciousness might work.
2. A good definition of 'genuine' randomness is the algorithmic one. A truly random sequence of numbers could not be generated by any computer program containing less information than that in the sequence itself. All programs for generating 'random' numbers contain a lot less information than the sequences they generate; therefore the sequences are pseudo-random. It's worth noting, though, that Stephen Wolfram has speculated that some of his 'cellular automata' may in fact behave in a 'truly' random manner, even though they are generated by quite short computer programs (see *A New Kind of Science)*. In that case, they would be random in the same sense as the digits of Pi.
3. By permission of Dean Radin and Simon & Schuster, Inc. From Dean Radin, *Entangled Minds,* Simon & Schuster, New York.
4. See note 3.

Further Reading

Chapter 1

Cohen J. and Stewart I. (1994) *The Collapse of Chaos: discovering simplicity in a complex world,* Penguin Books, London.

Desmond A. (1994) *Huxley: from devil's disciple to evolution's high priest,* Penguin Books, London.

Freeman W. (1999) *How Brains Make Up Their Minds,* Weidenfeld and Nicolson, London.

Jung C.G. (1976) *Letters.* (Ed: G. Adler), Routledge & Kegan Paul, London.

Kenet T., Arieli A., Tsodyks M. and Grinvald A. (2006) 'Are single cortical neurons soloists or are they obedient members of a huge orchestra?' In: *23 Problems in Systems Neuroscience,* (Eds: Leo van Hemmen and Terrence Sejnowski), OUP, Oxford.

MacCormac E. and Stamenov M. (1996) *Fractals of Brain, Fractals of Mind,* John Benjamins, Amsterdam.

Samuels A. (1985) *Jung and the Post-Jungians,* Routledge & Kegan Paul, London.

Senior C., Russell T. and Gazzaniga M. (Eds) (2006) *Methods in Mind,* The MIT Press, Cambridge MA.

Skarda C. and Freeman W. (1987) 'How Brains Make Chaos in Order to Make Sense of the World,' *Behavioural and Brain Sciences,* 10, 161–95.

Stevens A. (1990) *On Jung,* Routledge, London.

Chapter 2

Hofstadter D.R. (1985) 'Mathematical chaos and strange attractors.' In: *Metamagical Themas,* Penguin, UK.

Chapter 3

Baars B.J. (1997) *In the Theatre of Consciousness: the workspace of the mind,* Oxford University Press.

Braude S.E. (1995) *First Person Plural: multiple personality and the philosophy of mind,* Rowman and Littlefield, Maryland and London.

Dennett D. (2005) *Sweet Dreams; philosophical obstacles to a science of consciousness,* MIT Press, Cambridge MA and London.

Goertzel B. (2006) *The Hidden Pattern: a patternist philosophy of mind,* BrownWater Press, Boca Raton.

Henningsen P. (2006) 'Networks of Attractor Networks in the Brain.' This model is being actively developed. The most recent version can be viewed at http://www.netofans.net

Jeong J., Kim D.J, Kim S.Y., Chae J.H., Go H.J. and Kim K.S. (2001) 'The Effect of Total Sleep Deprivation on the Dimensional Complexity of the Waking EEG. *Sleep,* March 15; 24(2), pp.197–202.

Marshall L., Helgottir H., Molle M. and Born J. 'Boosting Slow Oscillations During Sleep Enhances Memory,' *Nature* 444, 610–13.

Nature Insight Review: Sleep (2005) *Nature,* 437, 1253–1289.

Panksepp J. (1998) 'The Periconscious Substrates of Consciousness,' *Journal of Consciousness Studies,* 5, no.5–6, pp.566–82.

Chapter 4

Arzy S., Idel M., Landis T. and Blanke O. (2005) 'Speaking with One's Self,' *Journal of Consciousness Studies,* 12, no.11, pp.4–29.

Bailey L.W. and Yates J. (1996) *The Near-Death Experience: a reader,* Routledge, London and New York.

Fox M. *Religion, Spirituality and the Near-Death Experience,* Routledge, London and New York.

Shanon B. (2002) *The Antipodes of the Mind: charting the phenomenology of the ayahuasca experience,* Oxford University Press.

Strassman M. (2001) *DMT: the spirit molecule,* Park Street Press, Rochester, Vermont.

Chapter 5

Gauld A. (1992) *A History of Hypnotism,* Cambridge University Press.

Schnabel J. (1995) *Dark White: aliens, abductions and the UFO obsession,* Penguin, UK.

Winter A. (1998) *Mesmerized: powers of the mind in Victorian Britain,* University of Chicago Press.

Chapter 6

Jung C.G. (1959) *Archetypes and the Collective Unconscious,* (Trans: F.C. Hull), Routledge and Kegan Paul, London.

Gray R.M. (1996) *Archetypal Explorations: an integrative approach to human behaviour,* Routledge, London and New York.

Chapter 7

Aunger R (Ed.) (2000) *Darwinizing Culture: the status of memetics as a science,* Oxford University Press.

Aunger R. (2002) *The Electric Meme; a new theory of how we think,* The Free Press, New York.

Blackmore S. (1999) *The Meme Machine,* Oxford University Press.

Jablonka E. and Lamb M.J. (2005) *Evolution in Four Dimensions: genetic, epigenetic, behavioural and symbolic variation in the history of life,* The MIT Press, Cambridge MA and London.

Nunn C.M.H. (1998) 'Archetypes and Memes: their structure, relationships and behaviour,' *Journal of Consciousness Studies,* 5, no.3. pp.344–54.

Chapter 8

Bartholomew R.E. (1994) 'Tarantism, dancing mania and demonopathy: the anthro-political aspects of "mass psychogenic illness",' *Psychological Medicine,* 24, pp.281–306.

Colombetti G. and Thompson E. (Eds.) (2005) 'Emotion Experience,' *Journal of Consciousness Studies,* 12, nos.8–10.

Hecker J.F.C. (1837) *The Dancing Mania of the Middle Ages,* 1885 edition, Humboldt Publishing Corporation, New York.

Wessely S. (1990) 'Old Wine in New Bottles; neurasthenia and ME,' *Psychological Medicine,* 20. pp.35–53.

Chapter 10

Agrippa H.C. (1530) *De Vanitate et Incertitudine Omnium Scientarum et Artium,* Translated and abridged by S. Brown, Amsterdam.

Barrow J.D. (1993) *Pi in the Sky: Counting, Thinking and Being,* Penguin, UK.

Edwards P. (1980) *Heresy and Authority in Medieval Europe: Documents in Translation,* Scholar Press, London.

Elliott B.B. (1962) *A History of English Advertising,* Business Publications Ltd, London.

Monter W. (1990) *Frontiers of Heresy: the Spanish Inquisition from the Basque Lands to Sicily,* Cambridge University Press.

Mullett M.A. (1980) *Radical Religious Movements in Early Modern Europe*, Allen & Unwin, London.

Penrose R. (2004) *The Road to Reality,* Jonathan Cape, London.

Pilditch J. (1989) *Winning Ways: How Companies Create the Products We All Want to Buy,* Mercury, London.

Plaidy J. (1959) *The Rise of the Spanish Inquisition,* Robert Hale & Co., London.

Sumption J. (1978) *The Albigensian Crusade,* Faber & Faber, London and Boston.

Chapter 11

Chandrasekhar S. (1987) *Truth and Beauty: aesthetics and motivations in science,* University of Chicago Press.

Claxton G. (1997) *Hare Brain, Tortoise Mind; why intelligence increases when you think less,* Fourth Estate, London.

Hadamard J. (1945) *The Psychology of Invention in the Mathematical Field,* Princeton University Press.

Livio M. (2002) *The Golden Ratio: the story of Phi, the extraordinary number of nature, art and beauty,* Headline, London.

Macfarlane R. (2003) *Mountains of the Mind: a history of a fascination,* Granta Books, London.

Ramachandran V.S. (2001) 'Sharpening up "The Science of Art",' *Journal of Consciousness Studies,* 8, no.1, pp.9–29.

Zeki S. (2002) 'Neural Concept Formation and Art,' *Journal of Consciousness Studies,* 9, no.3, pp.53–76.

Chapter 12

Cropper M. (2003) *The Life of Evelyn Underhill,* Skylight Paths Publishing, Woodstock, VT.

Fanning S. (2001) *Mystics of the Christian Tradition,* Routledge, London and New York.

Marshall P. (2005) *Mystical encounters with the Natural World: experiences and explanations,* Oxford University Press.

Newburg A.B. and D'Aquili E.G. (2000) 'The Neuropsychology of religious and Spiritual Experience,' *Journal of Consciousness Studies,* 7, no.11–12, pp.251–66.

Chapter 13

Barber J. (1999) *The End of Time: the next revolution in our understanding of the universe,* Weidenfeld and Nicolson, London

Brown J.W. (2000) *Mind and Nature: essays on time and subjectivity,* Whurr Publishers, London and Philadelphia.

Butterfield J. (ed.) (1999) *The Arguments of Time,* Oxford University Press.

Le Poidevin R. (2003) *Travels in Four Dimensions; the enigmas of space and time,* Oxford University Press.

Lockwood M. (2005) *The Labyrinth of Time: introducing the universe,* Oxford University Press.

Mellor D.H. (1998) *Real Time II,* Routledge, London and New York.

Price H, (1996) *Time's Arrow and Archimedes' Point: new directions for the physics of time,* Oxford University Press.

Susskind L. and Lindesay J. (2005) *An Introduction to Black Holes, Information and the String Theory Revolution: the holographic universe,* World Scientific, New Jersey, etc,.

Chapter 14

Buchanan M. (2000) *Ubiquity; the science of history . . . or why the world is simpler than we think,* Phoenix, London.

Healy D. (2002) *The Creation of Psychopharmacology,* Harvard University Press, Cambridge (Mass.) and London.

Smolin L. (2006) *The Trouble with Physics* (Chapter 16), Houghton Mifflin, Boston and New York.

Chapter 15

Grof S. (1993) *The Holotropic Mind,* HarperCollins, New York.

King C. (2006) 'Quantum Cosmology and the Hard Problem of the Conscious Brain.' In: *The Emerging Physics of Consciousness,* (Ed: Jack Tuszynski), Springer Verlag, Berlin.

MacCoun R.J. and Reuter P. (2001) *Drug War Heresies: learning from other vices, times and places,* Cambridge University Press.

'Psi Wars' (2003) *Journal of Consciousness Studies,* 10, no.6–7.

Radin, D. (2006) *Entangled Minds: extrasensory experiences in a quantum reality,* Paraview Pocket Books, New York.

Index